Farming and Profitability for Small Scale Farmers

Farming and Profitability for Small Scale Farmers

Nekesah T. Wafullah and
Kuria C. Munene

Delve
PUBLISHING

www.delvepublishing.com

Farming and Profitability for Small Scale Farmers

Nekesah T. Wafullah and Kuria C. Munene

Delve Publishing
224 Shoreacres Road
Burlington, ON L7L 2H2
Canada
www.delvepublishing.com
Email: orders@arclereducation.com

ISBN: 978-1-77469-089-5 (Hardcover)

ABOUT THE AUTHORS

Nekesah T. Wafullah is a skilled agriculture expert with extensive knowledge in agricultural energy value addition products, agricultural business management services, project management, various forms of fertilizer, their production, sales, marketing aspects and application regimes; cross border fertilizer trade policies; youth and women empowerment and volunteerism. She is adept at project planning and management as well as creating simple solutions to complex problems. She has experience within agricultural markets in Kenya, Zambia, Malawi, Tanzania, Rwanda, Democratic Republic of Congo (DRC)- Bukavu and Lubumbashi and Uganda. She mentors high school and college students and advocates for better performance in Agricultural science. During her free time, she loves editing books, watching movies, cooking, baking, networking, reading, and dancing. Nekesah holds an M Sc. In Agricultural and Applied Economics degree from the University of Nairobi with a major in International Trade and Policy.

Kuria C. Muene is an agriculture graduate with a masters degree in Agricultural and applied economics. He has extensive knowledge in agricultural research, policy analysis, supply chains, project management and crop management. He has experience in the use of data analysis tools that led to high-quality inferences for the betterment of the agriculture sector. He is an avid reader and an outdoor sports enthusiast in his spare time.

TABLE OF CONTENTS

List of Figures .... *ix*
List of Tables .... *xi*
List of Abbreviations .... *xiii*
Acknowledgement .... *xv*
Preface .... *xvii*

Chapter 1 **Small Scale Farming** **1**
Overview 2
Definitions, Characteristics and Contexts 3
Small-Scale Agriculture and Food Supply 4
Poverty, Hunger and Livelihoods in Small-Scale Agriculture 5
Small-Scale Farming Nature in the Us 5
Characteristics of Small-Scale Farming 7
Advantages of Small-Scale Farming 8
Contribution of Small-Scale Farming 9
Small-Scale Farming Challenges and Constraints 12
Factors Affecting Profitability of Small-Scale Farmers 15
Small-Scale Agriculture Dynamics 21
Transforming Small-Scale Farming Through Policy 26

Chapter 2 **Inclusive Value Chains** **37**
Overview 38
Agricultural Value Chains and Smallholders 39
Value Chain Models 44
Household Economy and Smallholders 50
Upgrading as Behavior Change 52
Including Small-Sale Farmers in the Value Chain 54
Impacts of Small-Scale Farmer Inclusion 56

Association and Inclusion of Small-Scale Farmers in the Value Chain....... 59
Governments, Donors and NGOS Initiatives in Developing Inclusive Value Chains 61

Chapter 3 Profitable Small Scale Farming Ideas........ 71
Introduction........ 72
Microgreens........ 73
Mushrooms........ 89
Soft Fruit / Fruit Trees Small Gardens........ 102
Micro Dairies........ 115
Small-Scale Chickens & Eggs Farming........ 121
Small-Scale Beekeeping........ 144
Fish Farming........ 154
Snail Farming........ 164

Chapter 4 Small-Scale Farming Profit Making Options........ 179
Overview........ 180
Small-Scale Machinery........ 181
Soil Health for Small-Scale Farmers........ 185
Permaculture........ 190
Forest Gardening........ 195
Aquaponics........ 199
Smart Water Management........ 215
Integrated Pest Management........ 224
Diversify the Smallholder Farm........ 227

References........ 231

Index........ 245

LIST OF FIGURES

Figure 1: General Value chain structure
Figure 2: Value Chain Map for Guatemalan High-Value Horticulture
Figure 3: Radishes
Figure 4: Broccoli
Figure 5: Peas
Figure 6: Mustard
Figure 7: Rocket
Figure 8: Life cycle of a mushroom
Figure 9: White Button (Agaricus bisporus)
Figure 10: Oyster Mushrooms (Pleurotus ostreatus)
Figure 11: Shiitake (Lentinus edodes)
Figure 12: Paddy straw (Volvariella volvacea)
Figure 13: Lion's mane (Hericium erinaceus)
Figure 14: Morel Mushrooms (Morchella esculenta)
Figure 15: strawberries
Figure 16: Raspberries
Figure 17: Blackberries
Figure 18: Currants
Figure 19: Blueberries
Figure 20: Gooseberries
Figure 21: Roman Snail (Helix pomatia)
Figure 22: vineyard snail (Otala lacteal)
Figure 23: White-lipped snail (Cepaea hortensis)
Figure 24: Giant African Land Snail (Achatina fulica)
Figure 25: 2 Wheel Tractor
Figure 26: Seven-layer system
Figure 27: Aquaponic system

LIST OF TABLES

Table 1: Profile of U.S. Small Farms

Table 2: Policy intervention for improving small-scale agriculture

Table 3: Proper policy intervention mix in promoting small-scale agriculture

Table 4: Key mushroom species and their corresponding cultivation medium

Table 5: Requirement of chickens for floor and perch space

Table 6: Predator attack modes and control methods

Table 7: Feed and feeder space requirements for 100 chickens

Table 8: The effect of creep feeding on flock egg production

Table 9: Minimum water and watering space requirements for 100 birds in hot dry conditions

Table 10: Characteristics of a good pond for fish farming

Table 11: Technology for Smart Water Management

Table 12: Diversification of agricultural systems

LIST OF ABBREVIATIONS

EPA	Environmental Protection Agency
CA	Conservation Agriculture
NFT	Nutrient Film Technique
PVC	Polyvinyl Chloride
OM	Organic Matter
SCPI	Sustainable Crop Production Intensification
RDI	Regulated Deficit Irrigation
SCADA	Supervisory Control and Data Acquisition
IPM	An Integrated Pest Management
DFS	Diversified Farming Systems
US	United States of America
SFRB	Scavenger Feed Resource Base
FAO	Food and Agriculture Organization
WWII	World War II
USDA	United States Department of Agriculture
LEDs	Light-Emitting Diodes
AI	Artificial Insemination
TMR	Partial Mixed Ration (PMR) Or A Total Mixed Ration
VC	Value Chain
SSA	Sub Saharan Africa
SPV	Special Purpose Vehicles
WBD	Wimm Bill Dann
CMRs	Calf Milk Replacers
HACCP	Hazard Analysis Critical Control Point
EPHEA	Ethiopian Horticulture Producers and Exporters Organization
VCD	Value Chain Development

NGO	Non-Governmental Organizations
IDH	Sustainable Trade Initiative
IFC	International Finance Corporation's
VCD	Value Chain Development
FDI	Foreign Direct Investment
FDA	Food and Drug Administration
DAFF	Department of Agriculture, Forestry, And Fisheries
SBA	Small Business Administration
RAI	Responsible Agriculture Investment, Such as The
WEF	World Economic Forum
WRI	World Resources Institute
IARAN	Interagency Research and Analysis Network
USAID	United States Agency for International Development
AMAP	Accelerated Microenterprise Advancement Project
BDS	Business Development Services

ACKNOWLEDGEMENT

The book is the product of great effort and time spent. The completion of this Book could not have been possible without the participation and assistance of so many people whose names may not all be enumerated. Their contributions are sincerely appreciated and gratefully acknowledged. However, I would like to express our deep appreciation and indebtedness particularly to the following: Arcler Education, Inc. and Anne Nanjekho for their endless support, kind and understanding spirit during this undertaking. To all relatives, friends, and others who in one way or another shared their support, either morally, financially, or physically, I would like to personally thank you.

PREFACE

A smallholder farm, is a small farm that follows the small-scale agriculture paradigm. Small-scale farmers are not all the same. Their circumstances, such as market access or environmental conditions, as well as their qualities, such as farm size, education, or financial holdings, are all extremely different. Agriculture on a small scale or by smallholders is a concept that is thrown about a lot. We've used the term small-scale farmer instead of smallholder in this paper because we're more concerned with the economic scale of a farming enterprise than with the size of the land. The argument that smallholders produce 70% of the food consumed by people living in LMICs is often used to justify the importance of small-scale agriculture. However, it has been demonstrated, that the story is more complicated. The final line is that a smaller number of "larger" small-scale farmers produce most of the food produced by small-scale agriculture. Small farms appear to meet the features of sustainable systems more closely than large farms, based on these lists of qualities (assuming the synonymity between large farms and industrialized ones). In agriculture, where the benefits of leveraging agricultural resources to the economic development process have been recognized for some time, the inclusive growth model has a significant appeal. More recently, USAID's Feed the Future Initiative underlined the importance of smallholder agriculture as a focal area in their strategy to achieving inclusive economic growth and improving food security, particularly in developing nations. An effective smallholder-led strategy for inclusive growth can promote a structural change of the agricultural sector by incorporating significant numbers of small-scale agricultural producers into competitive value chains. Increasing smallholder incomes, as well as multiplier effects on employment and income throughout the rural economy, lowering poverty, boosting food security, and fueling a process of local and national economic growth, are all advantages of such a structural transformation. Upgrading, which is measured in terms of productivity and efficiency, can help smallholders get access to higher-value markets. When it comes to value chains, the phrase "inclusive" emphasizes the importance of advantages for microenterprises, smallholders, and the low-income individuals and households who run them. An inclusive market system, in general, is one that benefits many low-income households. The whole range of activities and services required to move a product or service from conception to sale in its final market is referred to as a value chain. Participating in contract farming with domestic processors, exporters, or overseas customers is a significant way for rural farm-households, particularly in developing countries, to gain from agri-food exports and enhanced value in the domestic and export sectors. The amount to which smallholder farmers are incorporated in contract-farming arrangements, as well as the influence that involvement in contract-farming has on their incomes and well-being, determines whether they partake in the benefits of trade. Apart from being included in the value chain small-scale farmers have different ways of increasing social and economic gain. A productive farm does not necessitate a large quantity of land. Micro farmers are increasingly looking for property in urban areas that lack open space, but converting a small plot of land into a profitable enterprise requires more than a green thumb and a passing interest in gardening. Farming technology is continuously evolving, and staying on top of the latest methods and equipment is critical if you want to succeed in a competitive industry. Successful micro farmers treat their land as a business venture, embracing cutting-edge farming equipment to increase production and profits. Micro farms are small-scale agricultural activities that require significantly less land than a conventional commercial or family farm often less than five acres and are frequently found in urban or suburban areas. Micro farmers are forced to be innovative about the crops they raise and the way they seek to earn a profit because of size and zoning restrictions, and they tend to focus on sustainability, seasonal crops, and specialized markets for their products. Micro farms, despite their small size, can generate a large amount of food per acre, which, when correctly managed, can result in large profits. Farming profit margins are increased using tactics like those used in other businesses around the country. Farmers strive to maximize work, cut costs, and increase revenue in order to keep as much profit as possible. There are a lot of ways farmers can use to enhance profit margins without affecting the quality of the products they bring to market, other from just raising pricing. Farmers can reduce product shipping expenses and enhance profit margins by selling to local community markets and grocery shops. In addition, a farmer can save money on payroll by having shorter delivery routes for his drivers. Because an increasing number of people are eager to lessen their carbon impact by purchasing locally grown products. A farmer can cut energy expenses and boost profit by investing in alternative energy alternatives, such as wind power. For small scale farmers practicing organic farming, conservation tillage leads to economic, societal, and environmental benefits.

Chapter

1

Small Scale Farming

CONTENTS

Overview.......... 2
Definitions, Characteristics and Contexts.......... 3
Small-Scale Agriculture and Food Supply.......... 4
Poverty, Hunger and Livelihoods in Small-Scale Agriculture.......... 5
Small-Scale Farming Nature in the Us.......... 5
Characteristics of Small-Scale Farming.......... 7
Advantages of Small-Scale Farming.......... 8
Contribution of Small-Scale Farming.......... 9
Small-Scale Farming Challenges and Constraints.......... 12
Factors Affecting Profitability of Small-Scale Farmers.......... 15
Small-Scale Agriculture Dynamics.......... 21
Transforming Small-Scale Farming Through Policy.......... 26

OVERVIEW

A smallholding, also known as a smallholder farm, is a small farm that follows the small-scale agriculture paradigm. Small-scale farmers are not all the same. Their circumstances, such as market access or environmental conditions, as well as their qualities, such as farm size, education, or financial holdings, are all extremely different. The well-being of farm families, as well as their economic viability and involvement in food systems, is shaped by this combination of context and features. Small-scale farmers, on the other hand, are frequently considered as a single group in a single set of circumstances in the literature and in policy discussions. The definition of a smallholder or small-scale farm varies greatly, depending on characteristics such as size, food production technique or technology, family engagement in work, and economic impact. Smallholdings are often farming with a combination of income crops and subsistence farming that support a single family. Smallholdings may not be self-sufficient as a country becomes wealthier, but they may be cherished for the rural lifestyle. Some of these smallholdings are becoming more economically viable as the sustainable food and local food movements grow in rich countries. In developing countries alone, there are an estimated 500 million smallholder farms that support about two billion people. Industrial agriculture, which seeks efficiency by increasing outputs, monoculture, consolidating land under large agricultural enterprises, and economies of scale, frequently clashes with small-scale agriculture. Smallholders are primarily reliant on some labor-intensive income crops, such as cocoa production in Ghana or Côte d'Ivoire; as of 2008, smallholders grew 90% of cocoa globally. Cocoa provides 60 to 90 percent of the income for these growers. Other crops, such as coffee, palm oil, and bananas, have similar supply chain tendencies. Small-scale agriculture can stimulate food system investment in small-holder farmers in other markets, improving food security. Some firms are now attempting to integrate smallholdings into their value chain by supplying seed, feed, or fertilizer to help them increase productivity. Smallholder farms often require fewer industrial inputs and can be an important way to improve food security and sustainable food systems in less-developed contexts, so addressing their productivity and financial sustainability is a global development priority, as measured by indicator 2.3 of Sustainable Development Goal 2. Furthermore, because agriculture has such a big impact on climate change, Project Drawdown identified "Sustainable Intensification for Smallholders" as a critical technique for mitigating climate change.

DEFINITIONS, CHARACTERISTICS AND CONTEXTS

Agriculture on a small scale or by smallholders is a concept that is thrown about a lot (HLPE 2013). We've used the term small-scale farmer instead of smallholder in this research because we're more concerned with the economic scale of a farming enterprise than with the size of the land. On the same piece of land, various crops provide drastically different results. Small-scale/smallholder agriculture is defined differently in different countries. Smallholder refers to farms of less than 2.0 hectares, according to the most generally used definition (HLPE 2013, IFAD 2013). In common language, however, the word is frequently applied to any farmer who is not large-scale and/or financially disadvantaged. When it comes to figuring out who produces how much of the world's food supply, distinguishing between different scales and parts of "small-scale" production becomes crucial (Christen and Anderson 2013). Family farming and corporate farming are two separate types of agriculture. Family farming is defined by the FAO as "a method of organizing agricultural, forestry, fisheries, pastoral, and aquaculture output that is managed and maintained by a family and is primarily reliant on family capital and labor, including both women's and men's labor." The farm and the family are intertwined, co-evolve, and perform economic, environmental, social, and cultural functions" (FAO 2013). Smallholders are a vast, persistent, and internally diversified group that resists unduly tight definitions and overlaps but is not identical to the category of family farmers, as Zimmerer (2018) points out. To better characterize the variability of small-scale farming, several frameworks have been employed. Berdegué and Escobar (2002) developed a two-dimensional matrix to differentiate between farm production conditions and farmer assets, resulting in three distinct groups: subsistence farmers, small investor farmers, and large-scale farmers. Smallholders are divided into four groups, according to Elbehri et al. (2013):

- Those who practice subsistence farming and hence lack access to or refuse to participate in markets;
- those with limited access to markets;
- those with frequent access to markets; and
- those entirely dedicated to commercial farming.

Dorward et al. (2009) identified three strategies pursued by the rural poor,

- stepping up,

- hanging in and
- stepping out.

DFID adopted this concept in their 2015 Conceptual Approach to Agriculture. Mangnus and Metz (2019) provide a helpful review of food security and small-scale agriculture definitions, criteria, and targeting. Understanding small-scale farmers' roles in food systems and developing properly targeted solutions require disaggregating them according to conceptual frameworks. However, there is currently no systematic use of such frameworks, nor is there any data available to provide a more sophisticated disaggregated knowledge of small-scale agriculture. The only method to look at small-scale agriculture in a disaggregated fashion is in terms of farm size. Lowder et al. (2016) did a thorough analysis of the literature and data on farm size distribution and farm numbers. The following conclusions are drawn from their findings:

- Globally there are at least 570 million farms
- Of these at least 500 million, approximately 90% are family farms.
- Approximately 475 million farms or 84% are less than 2 ha
- While family farms operate about 75% of agricultural land the 475 million farms of 2 ha or less only operate about 12% of agricultural land.

SMALL-SCALE AGRICULTURE AND FOOD SUPPLY

The argument that smallholders produce 70% of the food consumed by people living in LMICs is often used to justify the importance of small-scale agriculture. However, as Ricciardi et al (2018) and Herrero et al (2017) demonstrate, the story is more complicated. The final line is that a smaller number of "larger" small-scale farmers produce most of the food produced by small-scale agriculture. The 72 percent of farmers with less than one hectare (410 million) are likely to produce less than 20 percent of the world food supply. The food produced by this bigger group of very small-scale farmers is important for their own food and nutrition security, as well as extremely localized markets, but it is insufficient to meet the expanding demands of urban populations. Meanwhile, the 14% of farmers who own between 2 and 20 hectares produce 31% of the food. Around 10% of the world's food is produced on farms of 20 to 50 hectares. While medium and small-scale farms of 50 ha provide 51-77 percent of food, depending on the country, the majority of this is produced by the 26 percent of farms of 1 – 50 ha, not the 72 percent of farms of 1 ha.

POVERTY, HUNGER AND LIVELIHOODS IN SMALL-SCALE AGRICULTURE

According to the worldwide poverty line, 736 million people, or 10% of the world's population, live in extreme poverty (FAO, 2018a). Rural areas are home to 80% of the severely poor (USD 1.25/day) and 76% of the moderately poor (Castaeda et al 2018). There's a strong link between extreme poverty and the 821 million people who are still hungry in 2018. Agriculture is the poor's primary source of income. 76.3 percent of the extreme poor and 60.7 percent of the moderate poor work in the sector (Castaeda et al 2018). Most of this group is subsistence or semi-subsistence focused, and they face severe obstacles to participating in higher-value agricultural operations. Not all farm households with less than one hectare are impoverished, but a significant number of them are. If one considers a family size of five, the 410 million farms correspond to a total population of 2.05 billion individuals, the vast majority of whom, if not impoverished, are unquestionably poor. When all farms with less than 20 hectares are included, the population rises to 2.79 billion. When landless farm labourers are included, the total population approaches 3 billion, or approximately 40% of the world's population. To put it another way, small-scale agriculture is responsible for the livelihoods of 40% of the world's population.

SMALL-SCALE FARMING NATURE IN THE US

Using the traditional definition of a small farm as one with yearly sales of less than $40,000, nearly seven out of ten farms in the United States fall into this category. While small farms account for only 10% of overall sales (about $16 billion in 1992 dollars), they control a third of the value of all farm assets (including 30 percent of all U.S. farm land). Minority-owned farms account for 2% of all small farms; 70% are handled by full owners, with the remainder by part owners and tenants. Thompson discovered the following in a thorough study of small farms in the United States:

- great geographic disparity,
- no typical small farm, and
- a great number of small farms that grossed well under $40,000 annually;

Small farms do, on average, spend more of their operational budget on purchased inputs like fertilizer, chemicals, and electricity, and far less on labor, than large farms do. However, Thompson (among others) links this to the fact that larger farms typically have labor requirements that family members alone cannot meet. True, the total number of farms in the United States has been steadily decreasing. Furthermore, the number of small farms has decreased. Between census years, the latter appears to have decreased by an average of 13%. On the other side, it appears that the number of mid-size and large farms (defined as those with yearly sales of $50,000 or more) has climbed in all but one census year, increasing by an average of 81 percent. When inflation is considered, however, the reduction in small farm numbers appears to be less significant; many farms that appear to be "mid-sized" (especially during the relatively high inflationary period of the 1970s and 1980s) are, in reality, "small" farms. While small farms' overall gross cash income is positive, significant, and growing, their net cash income is negative and declining rapidly. However, because many small farms rely largely on off-farm revenue, experience suggests that a lack of aggregate net cash income does not appear to be a threat to their collective survival. A significant goal of small farm operators, according to Glover and Kusterer, is to raise their families' "stability and income while maintaining their independence as owners and operators of a farm operation." Even if monetary returns are insufficient to pay overall costs, this purpose provides incentive that may assure their long-term economic viability. Small farm data are more difficult to get by around the world. "About half of the world's population is dependent on subsistence farming, about 4090 of total cultivated land is worked by small farmers, 60 percent of all farmers are small, and they account for less than 40 percent of all agricultural output," according to Wharton (as cited in Valdes, Scobie, and Dillon, p. 168). In most developing countries, peasants are reported to produce a "significant part" of fundamental food crops (Valdes, Scobie, and Dillon). Glover and Kusterer discovered, contrary to expectations, that global agribusiness growth has not eliminated small farms and is unlikely to do so soon, in a study of contract farming. Agribusinesses deal with small farmers for a variety of reasons, one of which is political. Small farmers' access to confessional credit systems and their proclivity to provide better materials when exceedingly high quality is required are two factors that drive this type of contracting.

Table 1: Profile of U.S. Small Farms

Item	Value In 1992	Percentage Relative to Large farms	Percent change in value 1982-92	Percent change in value 1987-92
Number (1,000)	1,386	66	–20	–13
Gross Cash Income (billion $)	19	10	7	6
Net Cash Income (billion $)	–0.6	—	–216	–134
Gov't. Payments (billion $)	1.1	12	48	–47
Assets (billion $)	296	34	3	–23
Debt (billion $)	35	25	–13	–20
Debt/Asset Ratio (average %)	11.6	72	–16	0
Minority-Operated (1 ,000)	34	—	N/A	–5
Part Owners and Tenants (1,000)	409	—	–16	–9

Source: USDA/Economic Research Service (1992)

CHARACTERISTICS OF SMALL-SCALE FARMING

- Land size: Small-scale farming only works on 1-10 acres, or even less in the case of city backyard farms. Some small farms are larger than 10 acres, but they are primarily focused on animal husbandry rather than farming crops for profit.
- When opposed to large-scale factory farms, small-scale farming is usually a more sustainable means of producing land. To save money and give better long-term stability, small scale farming frequently employs numerous approaches employed in sustainable farming.
- According to the USDA, small farm income ranges from $1,000 to $25,000. Most family-run farms operate on a tight budget, forcing them to optimize in every manner possible, such as by avoiding pricey technology and equipment.
- As a result, small-scale farms typically require a lot of human effort. Tractors and other machinery are rarely utilized around the fields, and most of the work is done by hand.

- When compared to traditional industrial farming, this method of farming can be described as an extremely effective technique of producing crops, typically exceeding production per land unit.
- Animals (especially larger ones) are uncommon on small-scale farms; instead, small farms mostly focus on food production, with hens and perhaps a pig thrown in for good measure.
- On small farms, families generally cultivate their own food and sell the surplus, whereas on larger farms, crops are planted solely for profit.
- Crop rotation techniques are commonly used on small farms, making them less susceptible to disease.
- Organic fertilizer is commonly used on small farms, while oil-based soil improvement is avoided.

ADVANTAGES OF SMALL-SCALE FARMING

While small-scale farming is a lot of effort, it also has a lot of advantages over traditional farming methods.

- Builds communities: A small farm provides fresh food to the local community, reuniting people with the food they eat. This contributes to the development of a community centered around a small farm, benefiting both the farm and the people.
- Improves community health: Introducing more nutritious and sustainable food to customers helps to improve their overall health. They will receive not only high-quality food, but also a greater number of locally grown seasonal foods.
- Creates jobs: During the grueling harvest season, small farms frequently seek outside assistance in exchange for money, lodging, and food. Harvesting, selling, and maintaining a farm is difficult work, and assistance is always welcome.
- Small farms increase soil quality: Soil health is rarely considered in industrial farming, but on small farms, the soil is the lifeblood of the operation and is thus treated with respect. Small farms frequently strive to not only preserve but also improve soil quality over time, so that future generations can benefit even more from this agricultural practice.
- Food security: small farms grow a wider range of crops. Larger farms frequently grow only a single crop on a huge scale, which

means that in bad years, large portions of the harvest can be lost. In those years, small farms helped to mitigate the food system's instability.

- Better for the environment: Most of the food produced on large farms travels over 1000 kilometers before reaching shop shelves. Furthermore, a lot of high-tech equipment is needed to create these massive amounts of food in the shortest amount of time. Pesticides and fertilizers based on oil also have a negative impact on the environment.
- More productive: Contrary to popular assumption, small farms are significantly more productive than larger enterprises when using unorthodox agricultural techniques. Increasing the number of individuals who have access to food that is not only more but also of higher quality.
- Can provide great quality at a reasonable price: Most small farms sell their products directly on the farm, avoiding transportation and other costs. If you opt to buy from a local farm, you will be able to watch your food being grown and will know where your money is going.

CONTRIBUTION OF SMALL-SCALE FARMING

Exploring the extent of societal "benefits" and "costs" of small farms as a group, at least qualitatively, is one way to analyze their contribution to sustainability goals.

Societal Benefits of Small Farms

Regardless of their name (little farmers, peasants, or subsistence farmers), they share numerous features with their larger counterparts that set them apart. These are some of them:

- the intensity of the man-nature relationship,
- the diversity of plant and animal life, and
- the diversity of income sources on small farms relative to large farms.

As a result, we can deduce that. Small farms, individually and collectively, could contribute to both economic and environmental sustainability if environmental sensitivity, biodiversity, and income diversity are among the fundamental criteria for sustainable agricultural development.

Matching the features of a sustainable system with those of small farms and large farms, respectively, is one technique to investigate the contribution of small farms to sustainability goals. Tisdell correlates the following traits with sustainable systems:

- the preservation of intergenerational economic prosperity,
- Human beings will continue to live indefinitely,
- Production and economic systems' long-term viability in terms of resilience and other characteristics,
- community sustainability, and
- conservation of biodiversity

Following Tisdell, characteristics of "modern" or "industrialized" agricultural systems include:

- high energy consumption;
- high chemical consumption;
- requiring intense management;
- emphasizing uniformity over variation in both products and environments; and
- appearing to rely on the outcomes of ongoing research to maintain production.

Small farms appear to meet the features of sustainable systems more closely than large farms, based on these lists of qualities (assuming the synonymity between large farms and industrialized ones). This matching of traits is summarized in the table. Within the framework of sustainability, Thompson highlights the following specific benefits of small farms:

Act as buffers against urban encroachment: According to Thompson, the number of small farms in a community is directly linked to its economic strength. This is because small farms, particularly in metropolitan areas, do not rely heavily on agricultural income and hence are not subject to the same economic pressures as bigger farms to sell property for development. Small farmers, it turns out, control a disproportionately large share of farmland in the eastern states, where urban pressures are also highest.

Provide scenic attributes: The aesthetic attractiveness of tiny family farms to tourists as they drive about is impossible to quantify, but it is undeniable. The "attraction" in Lancaster County, Pennsylvania, according to Thompson, is the enormous concentration of Amish and Mennonite farms. He goes on to say that this is the most productive farm county east of the

Mississippi, with annual farm gross sales of $700 million and annual tourist gross sales of $250 million. Thompson also sees small farms as "goodwill ambassadors" for the agricultural sector to the general public, thanks to the roadside markets that many of them operate.

Lower intensity of land use: Small farms use their land less intensively than large farms, which could be less harmful to the environment. Small farms, for example, devote more land to woodland (17 percent versus 5%); less cultivable land is actually cultivated and harvested on small farms (50 percent versus 80 percent); a greater percentage of cropland is used for pasturing livestock (3 percent versus 8 percent); and small farms maintain nearly twice as much cropland (despite accounting for only 2% of total cropland) (Thompson), Small farms have a higher level of "site specificity" and are more aware of the unique characteristics of the area in which they operate.

Greater reliance on conservation practices: Thompson brings up an intriguing contrast. On the one hand, a consequence of small farmers' lower land use intensity is that they may contribute less to soil erosion than bigger businesses. For example, a 13-year study of soil degradation between small-holder farming and large-scale irrigated land in Nigeria (Eussiet) found that soil degradation was more severe on the large farms, both qualitatively and statistically. On the other hand, it's possible that small farm agriculture is more fundamentally erodible in the aggregate, necessitating its management in pasture or cover crops to prevent erosion. Overall, Thompson contends that small farmers "must be better land stewards than their larger counterparts" since small farmers are less reliant on erosive row crops, farm fewer acres, and can dedicate more time to care for them. 4 The second attribute is in line with the fact that sustainable systems are "management-intensive."

Intergenerational transfer of practices: Small farms, indigenous peoples, and so-called "developing" countries are more linked with this factor. In general, these approaches rely more on natural processes and are therefore more environmentally friendly.

Societal Costs of Small Farms

Although it is sometimes said that small farms are less efficient and slower to adopt new technology than larger farms, when evaluated from a social viewpoint, these perceived costs may really represent benefits.

Inefficiency: In farming, it is widely acknowledged that economies of scale accrue. However, the economies of scale are exaggerated to the extent that environmental and other social costs are excluded from traditional efficiency measures. Then there's the argument that large, commercial farms are more "efficient" than their smaller counterparts because commercial farms make decisions primarily based on financial considerations. As a result, they would not use more than the "necessary" quantities of fertilizers and pesticides, for example. Even if this were true, it does not guarantee the long-term viability of commercial farms. Indeed, if the short-term and cumulative environmental consequences of management decisions are ignored, such operations will almost certainly become unsustainable in the long run. Small businesses that ignore social production costs would be guilty of the same thing, albeit on a smaller scale. The buffering effect associated with off-farm income can, in fact, be viewed as an advantage that would make small farms more competitive while contributing to food supply and keeping food costs to society lower than they would be otherwise.

Slower adoption of technologies: Because large farmers have easier access to capital, they are often the first to adopt new technologies (Bieri, de Janvry, and Schmitz) (Price). While technological use can be economically beneficial—especially in the short run for early adopters—technical change can be detrimental to society when distortions such as externalities are present. As a result, slower/lower technology adoption by small farmers as a group may really be a societal advantage.

SMALL-SCALE FARMING CHALLENGES AND CONSTRAINTS

Agriculture continues to rely heavily on smallholder farming. Climate change, a lack of sufficient agricultural infrastructure, a scarcity of farming skills, high levels of soil degradation, and difficult economic conditions are all obstacles to small-scale agricultural output. The following are some of the limitations of small-scale farming:

Unsustainable farming practices: Intensive tillage-based farming techniques that pay little or no attention to managing the soils and landscapes as part of a live biological and ecosystem resource base have resulted in the deterioration of agricultural soils and, as a result, a loss in their health and productive capability. In most rural farming communities, animals are permitted to graze on crop wastes without sufficient supervision or management during the winter. As a result, most agricultural soils in these cropping systems have low organic matter content, poor structure, and fertility (Dube et al., 2012). Farmers' efforts to reverse the condition, such as the rapid halt of soil surface disturbance (tillage), the retention of an organic (crop residue) soil cover or mulch, and the inclusion of a diverse crop mix, will exacerbate the situation. These techniques will nourish soil microbes, increase rainfall infiltration, and protect the soil from water and wind erosion (Kassam et al., 2010).

Climate change: Climate change, which is primarily caused by global warming, is a major environmental challenge that humanity is currently facing. In many parts of the world, it has been identified as one of the major causes of decreased agricultural productivity. There is sufficient evidence that climate change has already impacted agricultural production in Sub-Saharan Africa in the second half of the twentieth century (Barrios et al.,2008). Most of the crop farming in SSA is rain-fed, making it vulnerable to weather fluctuations and variations (Kurukulasuriya & Rosenthal, 2003). Furthermore, land resilience has decreased as a result of increasing land degradation, exacerbating the effects of drought on already vulnerable farming communities. South Africa may need to import up to 7 million tons of maize in 2016, roughly half of its total requirements, according to reports. High food prices and severe food shortages have also resulted from the drought, which have primarily affected rural households in poor communities.

Poor infrastructure and logistics: In most SSA countries, poor infrastructure continues to obstruct agricultural activities. The lack of agricultural infrastructure, such as fencing and equipment, as well as inadequate and deteriorated market facilities and transportation systems, such as road and rail, are the main obstacles. The most critical infrastructural constraint facing agricultural development is the road transportation system, which is the most vital for market development in terms of distribution of inputs and outputs to and from farms. Smallholder farmers rely on inefficient modes of transportation, such as the usage of animals, as a result of the weak rural road network. Underdeveloped roads and other critical physical infrastructure frequently result in high transportation costs for agricultural products to market and farm inputs, lowering farmers' competitiveness. Furthermore, energy in rural regions is expensive and frequently unavailable, limiting investment in cold storage facilities, irrigation systems, and farm processing equipment. Perishable goods such as dairy and vegetables are limited in their marketability due to a lack of storage and processing facilities. Trading is hampered by these infrastructure and logistical obstacles.

Education: Agriculture is seen as a critical tool for addressing poverty and unemployment (Department of Agriculture, Forestry, and Fisheries (DAFF), 2009). However, the agricultural sector in South Africa is not only dualistic, with a developed commercial farming sector coexisting with many subsistence (communally owned) farms, but it is still primarily in the hands of white commercial farmers in terms of actual production, education, and technological know-how (DAFF, 2009). As a result, the country's challenge is to integrate the historically marginalized black farming population into the mainstream agricultural sector through education and knowledge. One significant path for redressing previous disparities is to equip black farmers with agricultural skills to manage the natural resource base in a sustainable manner. Higher economic growth in South Africa will be impossible without tackling issues such as illiteracy and poor education levels, which are most widespread in rural regions, and where agriculture is most likely to play a key role in resolving both economic and human development challenges (OECD, 2006). Apart from natural resource issues, South Africa faces significant hurdles in terms of human capital (skills, knowledge, and health). The lack of skills is considered as a major impediment to economic progress, social development, and long-term employment growth. As a result, increasing human capital in agriculture is particularly vital, as a scarcity of skilled human resources is a key stumbling block to progress. Farmers in South Africa can benefit from education and training on the sustainable use of agricultural natural resources, which can help them contribute productively to the agricultural economy. However, the introduction of appropriate technology that will effectively contribute to enhanced food security, sustainable agricultural output, and rural development is one of the major concerns in the twenty-first century (Olubode-Awosola & Van Schalkwyk, 2006).

Limited access to production inputs and equipment: When it comes to getting agricultural inputs, most small-scale farmers have difficulties. Farmers lack access to high-quality, vigorous seeds (hybrids vs. open pollenated varieties), fertilizers, implements (tractors and planters), and other farm machinery (e.g., knapsack and boom sprayers). Lack of access to such inputs and equipment frequently leads to poor seed cultivar selection, fertilizer type and quantity application, and crop planting delays. Crop yields will be negatively affected as a result of this.

Access to markets and credit: Improved access to input and output markets is a critical need for the agricultural sector's transformation from subsistence to commercial production. Smallholder farmers must have access to efficient markets and local value-adding, as well as increased competition. Markets, on the other hand, are frequently hampered by insufficient property rights and high transaction costs (Hall & Aliber, 2010). Despite these obstacles, some small-scale farmers have been able to provide food for both themselves and the market (Ortmann & King, 2006). In most situations, however, rising farmers are unable to meet the retailer's requirements (without major support). Requirements like pack houses, cool rooms, full traceability, and product, soil, and water analyses come at a high cost that some commercial farms, much alone new farmers, can't afford. Poor education and illiteracy are additional major factors that would prohibit these farmers from meeting the merchants' record-keeping and paperwork standards. Credit is one of the most important sources of capital accumulation, and it can be seen as a tool for increasing production efficiency and income. Small-scale farmers in South Africa's communal areas, however, have limited access to inexpensive loans. These farmers' limited access to agriculture and support services is linked to socioeconomic issues, as well as the challenges they face with financial and other institutions due to high risk and transaction costs (Okurut et al., 2004).

FACTORS AFFECTING PROFITABILITY OF SMALL-SCALE FARMERS

There are several elements that influence a company's performance and profitability, as well as its long-term viability and sustainability. The following are some of the factors that are defined in UTA:

Demographics

Age, gender, ethnicity, and farming experience are only a few examples of demographics. According to Ahearn (2013), elder farmers tend to run larger farms, and younger farmers run smaller farms. This is most likely due to rising land prices and dwindling land supply. Older farms may also be more successful, according to USDA census data, because they have more assets and have paid down their debt, leaving more profit at the end of the day (USDA, 2009). They can also profit from economies of scale in procuring supplies or producing items with cheaper overhead per unit by operating larger farms. According to other data, younger farmers are more successful because they are more willing to adopt new technologies and research (Hoppe and Korb, 2006). Young farmers can apply best practices in order to boost productivity/quality and thus profitability on a smaller scale as a result of this. Farm exits are also influenced by gender and ethnicity, with women and black farmers being more likely to leave than their white counterparts (Hoppe and Korb, 2006). While the exact cause for departing is unknown, it is safe to presume that they are not as successful or profitable as their counterparts for whatever reason, and are motivated to leave due to financial concerns. Since small farms and ranches are sometimes classified as small enterprises, and research shows that gender plays a role in a company's capacity to thrive, similar issues could be at play in a lack of success and profitability. Less aggressive or risk-averse conduct among women in entrepreneurial scenarios, societal circumstances, or inability to obtain the necessary start-up or operator loans due to social/financial position, among other factors, have been suggested. Furthermore, one may extrapolate that farming demands a large degree of hard effort, which could be a barrier to admission, as women may prefer less physically demanding occupations to farming.

Farm/Land Characteristics

Larger enterprises are often seen to be better able to take advantage of economies of scale, which small-scale, diversified farms are unable to do. One of the most difficult tasks for young farmers and ranchers is locating good and/or economical farm land for purchase or rent, particularly farm land that is eligible for government assistance or subsidies (Ahearn, 2013). Despite the 2008 financial crisis and a general decline in real estate valuation, speculation in farmland and undeveloped land has not lessened. Increased bioenergy crop use (Mueller, Anderson, and Wallington, 2011) and investors buying farms as part of a low-risk portfolio are two reasons for this (Waggle and Johnson, 2009). Size, location (proximity or access to appropriate markets), and other considerations all play a role in the overall success of the enterprise, although agricultural land prices in the urban/rural interface are steadily growing. Beginning farmers (those who have been farming for less than ten years) operate farms of all sizes, but they are on average smaller than established farms, according to USDA data. Although many beginner farmers own some land, they have a higher proportion of rented land than experienced farmers (Ahearn, 2013). Debt from land acquisitions and rental charges can eat into profits and have a negative impact on sustainability and lifespan if operators can't keep up with the payments. Following the discovery of adequate land, the availability of water becomes the next key worry. Water rights are not usually transferred with the land, and the rights that are transferred may not be sufficient for any type of crop. Well permits can be difficult to obtain, and water quality can also be an issue, all of which create hurdles to entry into the market. The quantity, quality, and method of irrigation available can have an impact on the crops or animals that can be supported on a farm. Water costs, irrigation pipelines, and water delivery systems can all influence profitability and success. Some systems may allow for increased or decreased output, as well as a change in the quality of the crop produced. When considered alongside other parameters such as area and distance to market, seasonal limits such as duration of growing season, first and last frost dates, and other constraints can have an impact on profitability (Ash et al., 2007). (Pendell et al., 2011; Martinez et al., 2010). If the growing and selling season is fewer than six months, the farm must reach sales levels that will allow it to survive the non-production months, or the operators may be forced to seek outside work to augment farm income. Using unheated tunnels or hoop houses, some farms have been highly effective in extending the growing season at a minimal cost (Wien, 2009).

Distance from key markets appears to have an impact on an operation's success and profitability, according to research (Martinez et al., 2010). The further one gets from a major city, the greater the distance one must travel to sell items. This cost must be factored into product pricing, which may make sales prices uncompetitive with those of other farmers. Those who live too close to urban centers, on the other hand, may incur the additional cost of owning assets (land) in densely populated areas, raising costs and pricing of commodities. Some far-flung growers have proven to be successful in marketing over long distances based solely on reputation; an example of such a reputation is that of the "Green River Melons," which are well-known and sought after at farmers' markets and roadside stands despite being grown far from the Salt Lake metropolitan area (Warchol, 2004).

Financials

There is significant worry about the likelihood of producers knowing whether they are lucrative and successful from year to year. Most people can tell if they're successful and profitable, but knowing more explicitly based on a debt-to-income ratio or a return on assets ratio can be more difficult for a producer who handles more on-farm activities and less accounting and financial planning (Mishra, El-Osta, and Shaik, 2010). Individual producers have various success criteria. A ten-acre organic peach farm, for example, may differ significantly from a 10,000-acre grass-fed cow enterprise. The costs of inputs and the values of outputs will differ significantly. While one producer may view the absence of the need for off-farm income as a metric of success, another may look at return on assets or return on livestock investment. Businesses frequently have numerous enterprises, with some earning a profit one year and others not; for example, extra milk may be sold at a loss in the dairy industry, but cheese-making or other value-added goods may readily compensate for these losses (Thistlewaite, 2012). While established farmers and beginning farmers earn roughly the same amount each year on average, beginning farmers earn a bigger part of their income from sources other than farming. Farm revenues account for a higher portion of established farmers' income (Ahearn, 2013). However, USDA research shows that for the first few years of a new enterprise, breaking even is often the best that can be hoped for (Ahearn, 2013).

Science & Technology Integration:

New entrants into an industry can help to create an environment that is conducive to the adoption of new technologies, as well as boost productivity by creating competition (Hoppe and Korb, 2006). Farm exits—and farm entries—appear to be comparable in agriculture, according to USDA study, and farm exits—and farm entries—might play a significant role in introducing innovations and productivity increases (Hoppe and Korb, 2006). Accounting or bookkeeping software, Internet sales, marketing, advertising, and day-to-day operations management are examples of new science and technology integration.

Day-to-Day Operations:

If used correctly, specialized outside support can assist a company in meeting its financial objectives (Thistlewaite, 2012). Using a part-time accountant, for example, may help the farmer to focus more on production and operations rather than worrying about a section of his business for which he is not best prepared or suited. In order to minimize costs, small business owners frequently try to be jacks-of-all-trades (Thistlewaite, 2012). Marketing service or onsite director/manager, farm manager, lawyer/attorney, accountant/bookkeeper, web developer/graphic artist, insurance agent, commercial real estate agent, mentor/coach, consultant(s), Chamber of Commerce activities, Small Business Administration (SBA), Small Business Development Centers, and Cooperative Extension work are some examples of commonly used value-added outside services. Beginning farmers are substantially less likely than established farms to use USDA programs because they are less likely to produce traditional agricultural commodities. Beginning farmers may not meet the eligibility standards for some USDA programs since they are oriented for commodities (Ahearn, 2013). Justifying the usage of such services necessitates balancing costs and benefits, and the services required may vary depending on the type of business. Involvement of family members in the procedure can both help and hinder it. If a spouse or family member is involved, supportive, and actively participating in a business, success is regarded to be more likely than if only one partner is involved (Muske et al., 2009). Having children or other family members who can assist with operations or decision-making is another factor that can contribute to success. Businesses, on the other hand, have been found to struggle when family members are placed in positions based solely on their status as family members, rather than on their talents or competence. According to research, combining personal and business funds might have a negative impact on the operation (Yilmazer and Schrank, 2010).

Suppliers

Reliability as well as number of suppliers can affect success and profitability of a corporation (Gupta, 2004). (Gupta, 2004). If suppliers of commodities needed to operate a business are inconsistent, this can impair production, planting, and eventually the success and profitability of an enterprise. The more suppliers available, the more options or access a firm has to those things that it could need (seed, fertilizer, stock animals, etc). (Seed, fertilizer, stock animals, etc.). However, using only 1 or 2 suppliers consistently can help with gross purchase discounts. Too many or even too few suppliers might negatively affect both success and profitability (Gupta, 2004). (Gupta, 2004).

Customers/Buyers

Business reputation, brand names, and other distinguishing characteristics add to a product's intrinsic value and can raise the price a producer can demand over the cost of production (Wirthgen, 2005). This boosts an operation's net profit and, as a result, its overall profitability. Consumers will buy from a particular grower or producer again and again if they believe the product is of higher quality or value. This implies that producers must consistently provide high-quality produce to their customers. The size of the customer base can have a significant impact on success: the more customers served, the less impact there will be if the market has a bad day or if one or two customers leave the area. Having too many customers and not being able to meet all their needs, on the other hand, can be detrimental to your business. Repeat customers vouch for a producer's product quality and reputation, and are more likely to recommend them to others (Thistlewaite, 2012). Word of mouth is one of the most effective forms of marketing and promotion available, especially in this age of abundance. Switching barriers also raise the cost of switching to a new provider for consumers. Consumers consider the cost/benefit ratio when deciding whether to keep a current relationship, according to Bennis (1964). Even if the consumer is dissatisfied with the service and/or product, the relationship will continue if the costs of switching outweigh the benefits. Loss of special treatment, risk perceptions based on lack of experience with another provider, search costs, the need to explain preferences, and current relationship sunk costs are all common costs.

Products/Product Mix:

It is widely thought and written that not only the appropriate product selection but also the right product mix may help a company succeed (Thistlewaite, 2012). A business is more likely to succeed if it is the only one of its kind or provides a unique product that isn't available anywhere else and is something consumers desire or need. Furthermore, having too many or too few items or product categories might influence sales and force the business owner to spread himself or herself too thin. Creating value-added products to use up, extend the life of, and store excess food, as well as using or selling culls or byproducts, can provide the producer with additional streams of money that might otherwise be considered a loss. This may enable for at least break-even or, in some cases, a bigger profit than if the product were sold through traditional channels. Another way to supplement your normal sales is to engage in agritourism. Agritourism is thought to raise community knowledge and support for local producers, resulting in more customers, as well as providing an additional source of income for the operator/producer. This not only generates a product to sell, but also an experience in which the customer may get more involved in the production of their meal.

Markets and Marketing Environment:

Difficult regulatory standards, overregulation, excessive scrutiny, or a complete lack of regulation can all have severe consequences for a company's success and profitability (Salatin, 2007). A producer is unlikely to succeed if regulatory criteria are impossible to achieve and significant fines for noncompliance are levied. However, if there is little or no oversight and the industry is plagued by food-borne illness outbreaks, profitability will suffer (Mozdoszka, 2004). Customer reputation is very directly linked to branding and labeling, which can have a significant impact on success and profitability. The cantaloupe industry in Colorado is a bad example of branding and labeling since the Rocky Ford growers' label was smeared owing to improper use of its labeling by an unrelated farm over ninety miles away, harming the reputation of that organization (Whitney, 2013). Thirty-three individuals died and many more were sickened as a result of listeria poisoning; the guilty farm is now bankrupt. Growers of Rocky Ford melon are still paying the price for an unrelated farm's use of their label to sell melons at a premium in the market. Brigham City peaches or Weeks Berries of Paradise are a sought-after and prized item at farmers' markets throughout Utah and a few neighboring states, demonstrating the value of labeling (Adams, 2011). Customers are motivated to buy this produce because of its perceived quality and a family heritage of buying it.

SMALL-SCALE AGRICULTURE DYNAMICS

This includes the dynamics of small-scale agriculture transformation in connection to market, livelihood, investment, and risk structural changes.

Market Revolution:

According to Reardon et al. (2019), there has been a "silent revolution" in food markets in low- and middle-income nations. This covers the rapid expansion of small and medium-sized businesses in a transitional food industry. Over the last 20-30 years, food markets in developing economies and around the world have changed dramatically. This transition is expected to accelerate in the next decades. Rural-urban supply linkages, for example, have grown dramatically. For Africa, Haggblade, Hazell, and Reardon (2010) forecast 600–800 percent growth over three decades. In Southeast Asia, it was almost 1000 percent in the same era, according to Reardon and Timmer (2014). Deep structural alterations in procurement, retailing, supply chain coordination, ownership, and power are driving these market changes. Market structure shifts from "traditional" to "transitional" to "contemporary" (Reardon et al 2019). Traditional markets are very localized, informal (no contracts), spot markets (cash-based immediate transaction), and have few, if any, formal rules. Food is carried from rural to urban regions over vast distances in transitional markets, causing elongation. There is a combination of labor- and capital-intensive technologies as a more complicated system of intermediaries emerges. Although quality criteria are being imposed, most transactions continue to be based on spot market relationships. The supermarket model exemplifies modern markets. As marketplaces get longer in space, there is typically supply chain consolidation and concentration (e.g., a small number of merchants who buy directly from big scale suppliers/producers), and spot markets give way to contracts and futures. There is a significant degree of capital intensification, and markets, private and public norms become commonplace. The major finding of Reardon et al. (2019) is that transitional markets dominate modern markets in low- and middle-income countries and will continue to do so for the foreseeable future. The demand created by the market revolution is mostly met by a group of small-scale family farmers. However, further research is needed to determine which farmers are supplying this demand in which circumstances. Making the premise that market expansion offers opportunities for individuals with a different asset base or set of circumstances than those now involved should be done with caution. Despite transportation, finance, and market distortions, this significant market change has occurred as a result of large-scale endogenous mechanisms. It has very little to do with donor-funded market development projects. There is no doubt that these marketplaces can be upgraded and improved in terms of efficiency and inclusion. The evidence of this market revolution, however, refutes the assumption that most small-scale farmers fail to connect with markets. As one gets farther down the economic pyramid of small farmers, small-scale farmer participation in this market revolution rapidly declines. The primary development challenge is how far down the economic pyramid can commercially viable small-scale farmers be created, at what cost, and under what conditions.

Emergent investment:

Low- and middle-income nations are experiencing significant domestic investments in the food and agriculture sector, primarily on a medium and small size, as a result of this market revolution. Foreign business investors, national governments, and development organizations make considerably fewer investments in food and agriculture than domestic private investors (Lowder, Carisma, and Skoet 2012). As "emergent farmers" (Jayne et al. 2016) and food sector entrepreneurs, there is rising evidence of salaried urban elites making significant investments back into agriculture. This investment has both benefits and drawbacks. It allows governments to meet rising urban food demand while also propelling the agri-food sector forward. However, domestic application of ideas of responsible agriculture investment, such as the CFS-RAI, has been restricted thus far. Domestic land grabs, poor environmental practices, and terrible labor conditions are all risks. According to Jayne et al. (2019), medium-scale farms in Sub-Saharan Africa are the fastest-growing part of the family farm industry, controlling more area than large-scale farms. Rather than existing small-scale farmers graduating to become larger and commercially viable, much of this expansion comes from investment by urban and rural "elites" (emerging farmers). This tendency has far-reaching ramifications for small-scale agriculture's transformation (Burke, Jayne, and Sitko 2019; Chamberlin and Jayne 2019). If new agricultural market opportunities are taken up on a large scale by emerging investors, it may drown out prospects for current small-scale farmers, undermining the development narrative of reducing poverty by assisting small-scale farmers in connecting to markets. The magnitude at which this emerging investment is taking place, as well as its impact on the commercialization of current small-scale farmers, are still unknown. However, in order to grasp the dynamics of converting small-scale agriculture, it is a trend that requires close attention.

Livelihood realities:

Living income and income diversification are two important features of livelihood reality. To afford good housing, nourishment, healthcare, education, and to deal with unforeseen or one-time expenses, a living income is required (Gneiting and Sonenshine 2018). Few small-scale farmers can make a living due to current agricultural pricing, tiny production scales, and low productivity. Global agribusiness firms have drawn attention to living income because they recognize that, despite efforts to include small-scale farmers in their supply chains on more equitable terms, their earnings still fall short of a living wage (see Farmer Income Lab and Living Income Community of Practice). This is considered as a barrier to long-term supply continuity, as well as a risk to the company's reputation. Many small-scale producers will remain at the bottom of the economic pyramid, below or just over the poverty line, until their terms of trade or other sources of revenue alter dramatically. It is not possible to solve this problem by increasing prices or yields by 20%, 30%, or even 100%. To get close to a living wage, they'd have to increase their revenue dramatically. Under current food pricing and policy conditions, graduating to commercially viable farming that provides a living income is not a realistic path for most of this group.

On- and off-farm employment, remittances, non-farm micro enterprises, trading, and social protection payments are all diversifying livelihoods for many small-scale farmers. This has three consequences. For starters, farming families are no longer as reliant on agricultural output and income as they once were, as evidenced by the high amounts of food purchases made by farming households. Second, the return on labor from farming relative to other employment possibilities becomes more relevant than the overall farm income. Having a small piece of land does not have to be a problem if it supplements other sources of income (and provides competitive/profitable labor returns). Third, access to alternative sources of income may reduce farmers' incentives to utilize their land fully or productively. This can affect a country's overall agricultural productivity if it affects many farmers.

Geographic and political economic context:

Farmers live in a variety of geographical and political economic environments. These distinctions are crucial in determining what restrictions and possibilities they confront, as well as which policies and interventions will be most effective. Understanding the nature of the transformation challenge requires disaggregating farming households in relation to their surroundings. We offer four scenarios: the first is long-term crisis and conflict zones. The second are marginal areas with insufficient connection and services such as roads, market facilities, energy, extension, and financial services, as well as constrained environmental conditions for agriculture. The third category includes places where environmental circumstances and connectivity are conducive to commercially focused small-scale agriculture, but this potential has yet to be fulfilled. The fourth scenario is where there is already a significant amount of small-scale and large-scale commercial agriculture, as well as well-developed market infrastructure and services. Areas of long-term crises and conflict may intersect with the other three contexts.

Farm characteristics:

Individual farming households have a wide range of factors that affect their commercial potential and poverty status in any scenario. Land size, economic activity scale, human capacity, financial assets, gender and age, family size, degree of family labor, resource tenure, and resource base quality are some of these factors. These traits, as well as the context in which they farm, must be considered when developing assistance strategies for small-scale farmers.

Risk:

Small-scale agriculture is increasingly vulnerable to a variety of threats. Market price fluctuations, weather variations, disease risks, natural disasters, poor quality inputs that result in low productivity, diminishing soil quality, and poor contract enforcement with exploitation by input suppliers, traders, and money lenders are all examples of these hazards. Extreme weather, disease, and natural disaster hazards are quickly increasing as a result of climate change. Agricultural markets are designed to shift risk down the value chain and onto producers. High levels of risk limit financial institutions' motivation to serve the agriculture industry and make it difficult for farmers to obtain financing. Poorer farmers are frequently forced to return to poverty as a result of climate and market conditions beyond their control. Adoption of new technologies and management methods is hampered by risk.

Small Scale Farming Future

In terms of markets, commerce, climate impacts, technology, socio-economic variables, and politics, the future of small-scale agriculture is uncertain. Any one of these variables has the potential to significantly upset current patterns, with positive or negative consequences for small-scale producers. As a result, the transformation of small-scale agriculture must be considered in terms of various future scenarios. There are two dimensions to this. One is looking into how food systems might shift and what it means for small-scale agriculture. The second is to look at different future possibilities to see how small-scale farmers' well-being and contribution might be maximized. Scenario thinking is a method of imagining how the future might unfold and what it might mean for various people and interests. Circumstances can aid in the clarification of desirable and unpleasant futures, as well as the development of intervention tactics that are resilient in all potential scenarios. Scenario analysis falls under the umbrella of foresight and future studies. We believe that conducting educated foresight and scenario analysis on the future of food systems and small-scale agriculture, especially at the national level, is a vital component of building knowledge and political will for change. The Food and Agriculture Organization (FAO 2018), the World Economic Forum (WEF) (2017), the World Resources Institute (WRI) (Searchinger et al. 2019), and the Interagency Research and Analysis Network (IARAN) have all recently conducted food systems scenario evaluations (2019). Three major scenarios can be outlined to assist focus thinking on more desired and less ideal futures. These are the following:

Business as usual: Despite attempts in many nations to achieve the SDGs, the future continues to follow trends that do not address fundamental concerns such as food access, availability, sustainability, and equity. Small farmers' condition improves marginally in terms of general food security and access to new markets, but climate change, unequal distribution of economic development, technical innovation, and investments have a significant impact.

Towards sustainability: Improvements in the socioeconomic and environmental sectors promote equitable and long-term access to basic human services, increase food security, and make major strides toward resource-efficient production and consumption. Income disparity, poverty, and hunger are reduced for small-scale farmers, who also get equitable access to basic services. Improved incentives and increased investments in sustainable practices and agricultural R&D make technological adoption simpler. Small-scale farmers play an important role in satisfying the evolving food needs of urban populations.

Stratified societies: Sustainability and equitable development are more difficult to achieve in a deeply divided global community. Elite classes safeguard their own interests and use their decision-making authority to promote unsustainable growth. As a result of protectionist measures, poverty and food insecurity are increasing, resource depletion is worsening, and regional and global fragmentation is increasing.

TRANSFORMING SMALL-SCALE FARMING THROUGH POLICY

Small-scale agriculture transformation must remain a key development goal in the next decades. This is to ensure that a huge number of people are not left behind, as well as to set food systems on a more sustainable and healthier path. New thinking, frames, and visions, on the other hand, are sorely needed. Future scenarios must be built on a fuller grasp of the structural changes occurring in agricultural systems and rural economies. It is critical to have a much more complex and disaggregated understanding of different types of farmers in various circumstances. Rural poverty and the challenging conditions faced by most small-scale family farms have become a "stuck" issue. Despite the attention the issue has gotten, major development investments, and multiple papers with similar recommendations, little has changed in terms of the size of rural poverty over the last decade. This is counterintuitive in some ways. During the same time period, the above-mentioned deep change in the dynamics of food markets has kept up with the demands of rapid population increase and urbanization. This halt in progress in converting small-scale farming is due to five interconnected reasons:

- The duality of small-scale agriculture is an oversimplification of the multiplicity of "small-scale farmers" and their various conditions and requirements.
- A lack of political will at all levels to develop the policies and make the public investments necessary to enable an inclusive and long-term transformation of small-scale agriculture (Birner 2018).
- Limited long-term and systemic guiding visions for the future of food systems and small-scale agriculture at the country level.
- Interdisciplinary and systems analysis, conversation, and engagement procedures in government, industry, civil society, and research are all lacking.
- Small-scale farmers have minimal market power and political clout.

In the end, small-scale agriculture reforms can only be fueled by a deeper national and local understanding, as well as transformative alliances that generate political will for policy reform and public investment. We believe that a desirable transformation of small-scale agriculture involves three interrelated factors from a systems perspective:

- Clarifying the desired objectives of small-scale agriculture transformation from a food system viewpoint, using scenario thinking to consider trade-offs, synergies, and paths;
- determining the mix of actions required to achieve desired results and evaluating their efficacy considering various future possibilities; and
- Putting in place effective mechanisms that combine science with policy, politics, and public debate and enable a more systematic approach to change through multi-stakeholder participation.

Technical inputs are required for these parts, but they are mostly about institutional and political innovation (Woodhill 2010). Small-scale agriculture's future transformation must be considered in the perspective of the food sector's and economy's fundamental structural changes.

Desired Small-Scale Agriculture Outcomes

Economic and social wellbeing:

- Small-scale farmers are part of a food system economy that maximizes fair employment for the greatest number of people.
- Small-scale farmers can support themselves.
- Off-farm work alternatives are available in vibrant rural economies.
- A planned exit from agriculture for those who no longer desire to farm and/or are unable to support themselves.
- Equal decision-making authority and access to resources/tenure are gained by women.
- Women are welcome and safe in agricultural marketing.
- Large-scale outmigration of destitute rural residents, which would exacerbate urban poverty, is avoided.

Food and nutrition security:

- Farm families eat a well-balanced diet.
- Small-scale agriculture helps urban residents eat a healthy, well-balanced diet.
- Food shortages/price shocks caused by market swings, climate change impacts, and natural disasters are mitigated by household savings schemes, social protection, and improved farming techniques.
- To increase food quality and safety, informal and semi-formal supply chains are being modernized.
- In order to ensure national food and nutrition security, an optimal balance between small-scale agriculture, larger-scale farming, and imports must be achieved.

Environmental sustainability:

- Small-scale farmers use climate-smart farming techniques to minimize emissions and build resilience to the effects of climate change.
- Small-scale agriculture is incorporated into sustainable landscape initiatives that combine farming with biodiversity conservation and ecosystem service protection.

- Small-scale farmers use best practices for soil and water conservation.
- Agrochemicals are used by small-scale farmers in a minimal/optimal and safe manner.
- Small-scale farmers can reduce food waste to a minimum.

Trade-Offs and Implications

Prices for consumers vs producers:

- **Implications:** A challenging policy issue is whether to increase farmer income and compensate for food's environmental costs, or to provide consumers, particularly those in cities, with cheaper and more readily accessible food. Because of the political difficulty of this choice, the agriculture sector receives substantial subsidies.

Number of producers vs size of landholding:

- **Implications:** Due to limited land resources in many places, land consolidation is required to boost agricultural scale and profitability. This creates significant trade-offs between the security of extremely tiny but marginal portions of land for many people versus a smaller number of commercially viable farmers. The nature of this trade-off is very dependent on alternate sources of income.

Larger vs smaller scale agricultural Production:

- **Implications:** Potential tradeoffs between labor and capital, economic efficiency vs. economic inclusion, and farmer resilience vs. vulnerability exist in larger vs. smaller-scale production systems.

Off-farm vs on-farm livelihoods:

- **Implications:** There are trade-offs between having a high number of rural poor people owning their own property and producing, even if only marginally, or having many people informally hired as farm laborers on extremely low and itinerant salaries.

Open trade vs protectionism:

- **Implications:** Open trade can expand market prospects, boost export revenue, and lower consumer prices. It can also make it difficult or impossible for farmers to compete with imports, halting the development of certain industries.

Investment in agriculture vs other sectors:

- **Implications:** For national governments, there are trade-offs between investing limited public funding in agriculture versus other industries. This all boils down to basic economic assumptions regarding agriculture's role in economic development.

Calories vs nutrition:

- **Implications:** Small-scale agriculture focused toward production of staple food crops to meet basic calorie demands vs. systems that produce a variety of nutrient varied crops differs dramatically. There are expenses associated with establishing more nutrient-beneficial systems against the costs associated with inadequate nutrition.

Domestic vs international food security:

- **Implications:** Countries might choose to achieve food security through self-sufficiency, international trade, or a combination of the two. The dangers of high global prices and a breakdown of open commerce are balanced against the hazards of a country's inability to meet its demands due to drought, natural disasters, or illness. In terms of costs and effects for domestic manufacturers, there are also trade-offs.

Subsidies vs alternative public investments:

- **Implications:** Subsidies for agricultural inputs or price support schemes can boost productivity in the short run, boosting food security and farmer earnings. However, there may be a trade-off between investing in infrastructure and services that will result in a more sustainable economy in the long run. Creating effective agricultural market systems might also be a tradeoff.

Formalized vs semi-formal value chains:

- **Implications:** Food markets in the West are highly structured, centralized, and employ a small number of people. With a significant number of small businesses and high levels of employment, emerging country markets remain semi-formalized. In terms of prospects for small-scale farmers, inclusive economic growth, food quality and safety, foreign investment, and export opportunities, the transition between the two entails trade-offs.

- **Policy Interventions:** Small-scale agriculture and food systems transformation are systemic issues. They necessitate solutions that consider the interactions of entire systems. Controlling and engineering the behavior of complex human and natural systems is impossible. Smart interventions that address system linkages, on the other hand, can nudge systems in more desired directions while dampening unwanted ones. This necessitates a continuous process of policy learning and modification. In the single sector, piecemeal, short-term, and powerful interest group driven processes that often characterize public policy development, radical reforms are required. Not that changing the trajectory of a business-as-usual scenario to something more desired for all is simple, but it will necessitate new and different ways. For such a comprehensive examination, the food systems framework serves as a starting point. It serves as a foundation for mapping out the relationships between major food system actors and the incentives that drive their activities, whether at the local level for a single value chain or at the national level for the entire food system. To better identify intervention areas and build theories of change to steer transformation, a fuller understanding of the food system's fundamental operations and how they interact with supporting services and the institutional environment is required. Rapidly evolving computer modeling and visualization capabilities bring up huge opportunity for more advanced analysis and hypothetical testing of what might work, as well as collaborative exploration with stakeholders.

There is no shortage of broad advice regarding what needs to be done to improve small-scale agriculture, as seen in the table below and research dating back more than a decade. Better public services, financial services, improved input and output market functioning, research and development, infrastructure, and successful producer organizations are all on the menu. What's needed are country-level and more localized small-scale agriculture transformation strategies with clear outcomes, as well as integrated, systemic change routes tailored to the needs of various small-scale farmers. Too often, ideas and policy orientations for enhancing small-scale agriculture fail to distinguish between the demands of various types of small-scale farmers in various circumstances (Vorley et al, 2012; Hazel and Rahman, 2014). Based on the transition methods of stepping-up, hanging-in, moving-out, and adding an extra level of stepping-in (emergent/investor farmers), Table

4 shows how differentiated policies could be tailored to distinct groups of small-scale farmers.

Table 2: Policy intervention for improving small-scale agriculture

Intervention	**Detail**
Good governance and enabling business environment	Building ownership, working toward responsible governance, decentralizing, and devolving policy and operations, designing, and implementing medium- and long-term strategies, implementing internal reforms, and encouraging the creation of enabling environments with built-in incentives for addressing market failures are just some of the things that need to be done.
Public services and infrastructure	Investing in public goods, services, and infrastructure in rural communities, expanding access to and affordability of smallholder-friendly technologies, and encouraging innovation.
Market failures	Supporting access to markets and market information, encouraging smallholder diversification, and focusing governmental interventions on market failures are all things that can be done.
Financial services	Improving access to innovative financial services, extending risk planning opportunities for small-scale farmers, and revising financial rules are all things that need to be done.
Research, development, and advisory services	Improving and increasing farm data collecting, investing in agricultural research setup, and upgrading, and providing funding for technology and agricultural innovation.
Aggregation and representation	Farm groups are being empowered via training, development, and modernization of commercial and technical skills.
Sector development and value chains	Increasing market prospects by promoting pro-smallholder value chains, vertical and horizontal integration, and the development of fledgling processing sectors.
Public private partnerships	Creating clear regulatory frameworks for connecting farmers with private institutions, as well as creating collaboration and facilitation methods.
Territorial/rural development	Developing policies that encourage and incentivize rural development, non-farm employment, and rural-urban connectivity, as well as encouraging context-specific policies.

Gender	Through inclusive decision-making and policy processes, increasing access to knowledge, technology, assets, and resources to boost employment prospects, minimize drudgery, and empower people.
Social protection	Creating and sustaining social safeguards, assisting with training and competency development, and combining productivity-enhancing techniques with social protection and social safety nets are all things that need to be done.
Land tenure and property rights security	Creating flexible and clear land transfer and tenure security arrangements, implementing land agricultural reforms with active assistance for small-scale farmers, and developing supporting regulatory and legal frameworks for land acquisitions.

Source: Woodhill et al., (2020)

Infrastructure, an enabling business environment, education and health services, and access to finance are among the policies that assist all types of farmers and the wider rural economy, as noted by Hazell and Rahman (2014), Wiggins (2014), and Mellor (2017). Economists have often stated that public funds should be invested in these forms of public goods and services rather than in input subsidies and price support schemes, which often cause market inefficiencies and are difficult to target to those who are most in need. The extreme impoverished and food insecure should therefore be targeted with social protection systems. Over the last decade or two, the demands of commercializing small-scale agriculture and connecting farmers to markets have received a lot of attention (see Woodhill 2016). While unquestionably vital, it is possible that this has diverted focus away from the requirements of the vast majority of (semi-)subsistence farmers who, even selling surplus into informal local markets, will never become commercial. This category, which is largely poor to severely poor, must maximize their productive and income-earning options and capacities while simultaneously maintaining their food and nutrition security. This necessitates innovative thinking on how to link social protection, rural development, and agricultural policies in a synergistic manner.

Table 3: Proper policy intervention mix in promoting small-scale agriculture

Small-Scale Farmer Category	Proper policy intervention
Stepping in	Incentivizing inclusive and responsible agri-food industry investment that does not stifle existing small-scale farmer possibilities
Stepping up	Enabling small-scale farmers to scale up, commercialize, be competitive and profitable, and manage risk by providing a business environment and access to services and technologies.
Hanging in	Increasing farm income and food security while connecting to productive safety nets, increasing access to off-farm income, and strengthening shock resilience.
Moving out	Off-farm employment education, transitional safety nets, equitable land transfer and renting, rural economic development, and enterprise development support

Source: Woodhill et al., (2020)

Even though interventions are required for all types of farmers, such as access to finance, extension services, women's empowerment, or property rights, tailoring interventions to the specific needs of certain groups and circumstances remains critical.

Recommendations for Small-Scale Agriculture Transformation

- **Convene a global learning platform on small-scale agriculture transformation.** In recent years, the global emphasis on the future of small-scale agriculture appears to have shifted away from small-scale agriculture. To keep the understanding of small-scale agriculture up to date, learn lessons from development interventions, and shape a future agenda that aligns with the emerging attention on food systems, more coordinated efforts across international agencies, donors, development organizations, regional organizations, and business are needed.
- **Invest in research, data, and synthesis.** Small-scale agriculture data is few and scattered, and there has been insufficient meta-analysis and synthesis of the findings of various studies. There is also a meta-analysis gap in terms of the magnitude of impact market-driven approaches to small-scale agriculture transformation have had, both in terms of the number of farmers who have benefited and the degree of improvement in livelihoods.

- **Undertake and collate case studies.** Case studies assessing large-scale and autonomous changes, as well as examples of positive deviance in small-scale agriculture, could provide more insight. Such examples, when evaluated against a variety of contexts and policy settings, could aid in the development of more evidence-based policy innovation.
- **Develop country level small-scale agriculture portraits.** These will enable tracking of trends and intervention targeting by providing a much clearer and more comprehensive categorization of small-scale farmers in terms of their context and characteristics.
- **Public policy and investment options.** While there is a vast list of interventions that can help small-scale farmers, country-specific solutions and scenarios must be developed. Such frameworks may be able to encourage more policy innovation.
- **Develop national level small-scale agriculture transformation strategies based on food systems foresight and scenario analysis.** In the end, countries will need to build more integrated policies to guide small-scale transformation and allow for the integration of suitable actions across sectors. These must include reasonable projections for the number of farmers who will achieve profitable commercialization, the number of farmers who will leave agriculture, and the number of farmers who will require transitional assistance.

Chapter 2

Inclusive Value Chains

CONTENTS

Overview..38
Agricultural Value Chains and Smallholders..39
Value Chain Models..44
Household Economy and Smallholders...50
Upgrading as Behavior Change..52
Including Small-Sale Farmers in the Value Chain.....................................54
Impacts of Small-Scale Farmer Inclusion..56
Association And Inclusion of Small-Scale Farmers in the Value Chain......59
Governments, Donors and Ngos Initiatives in Developing Inclusive Value Chains...61

OVERVIEW

In agriculture, where the benefits of leveraging agricultural resources to the economic development process have been recognized for some time, the inclusive growth model has a significant appeal (Johnston and Kilby 1975, Mellor 1976). More recently, USAID's Feed the Future Initiative underlined the importance of smallholder agriculture as a focal area in their strategy to achieving inclusive economic growth and improving food security, particularly in developing nations. An effective smallholder-led strategy for inclusive growth can promote a structural change of the agricultural sector by incorporating significant numbers of small-scale agricultural producers into competitive value chains. Increasing smallholder incomes, as well as multiplier effects on employment and income throughout the rural economy, lowering poverty, boosting food security, and fueling a process of local and national economic growth, are all advantages of such a structural transformation. Upgrading, which is measured in terms of productivity and efficiency, can help smallholders get access to higher-value markets. When smallholders invest in increasing their own productivity, it can help the entire value chain compete more effectively. Smallholders become more appealing commercial partners to larger enterprises in the value chain as their productivity rises and their capacity to react to changing market conditions improves. While upgrading (by definition) adds value, smallholders must capture some of that value in order to reap the benefits in the form of increased household income. Upgrades for smallholders are a set of investment decisions made at the intersection of market and home systems. Smallholder households provide productive land, labor, entrepreneurship, and other rural resources, whereas market systems provide demand-driven opportunities based on profit incentives. Because smallholder households produce and consume food, their farming decisions can have an impact on both income and food security. Furthermore, household income diversification can mean that smallholders allocate their resources among a variety of productive activities, such as food crops, cash crops, livestock, off-farm work, and rural microenterprises. It's vital to recall that smallholders arrange their resources among competing investment, output, and consumption options in order to understand their upgrading behavior. Smallholder engagement in agricultural value chains, smallholder upgrading behavior, and economic effects relating to agricultural production and smallholder incomes are all discussed in this paper. Here are some key definitions to consider in this discussion:

- **Inclusion:** How involved have smallholders been in agricultural value chain projects? What are the different sorts of smallholder outreach projects?
- **Upgrading:** Have smallholders been willing and able to upgrade in order to increase value? What types of productivity effects have been observed in which circumstances?
- **Benefits:** Is it possible for smallholders to recoup part of the increased value created by upgrading? Do smallholders and their families gain financially from their participation in and upgrades to their properties?

AGRICULTURAL VALUE CHAINS AND SMALLHOLDERS

From their position at the crossroads of various systems, including agricultural value chains and farm family economies, smallholders make decisions to enter new markets and invest in upgrading.

Value Chain Approach and Framework

In recent years, the United States Agency for International Development (USAID) and other international development agencies have incorporated concepts from the global value chain literature (Gereffi and Lee 2012) to construct a holistic approach to attaining inclusive economic growth (USAID 2012). The Accelerated Microenterprise Advancement Project-Business Development Services (AMAP BDS Project) (1994-2012) developed the inclusive value chain approach—a private sector development approach designed to achieve the dual objectives of economic growth and poverty reduction—under the leadership of USAID's Microenterprise Development team. The value chain strategy is part of a larger trend in USAID's economic programming to help people participate in more inclusive market systems. When it comes to value chains, the phrase "inclusive" emphasizes the importance of advantages for microenterprises, smallholders, and the low-income individuals and households who run them. An inclusive market system, in general, is one that benefits a large number of low-income households. The whole range of activities and services required to move a product or service from conception to sale in its final market is referred to as a value chain. The structural and dynamic elements of the system are included in the conceptual framework for studying value chains. All the firms and other actors participating in the value chain, the vertical and horizontal links between these actors, supporting markets, and the business enabling environments within which actors make their decisions are all illustrated in the overall value chain map. The nature of the relationships between enterprises, as well as the value chain's ability to remain competitive by upgrading in response to changing end market demands and requirements, are dynamic aspects in the value chain framework. This section delves deeper into these structural and dynamic components. A value chain's structure encompasses all the involved firms and other agencies, and it can be described in terms of the five structural aspects listed below:

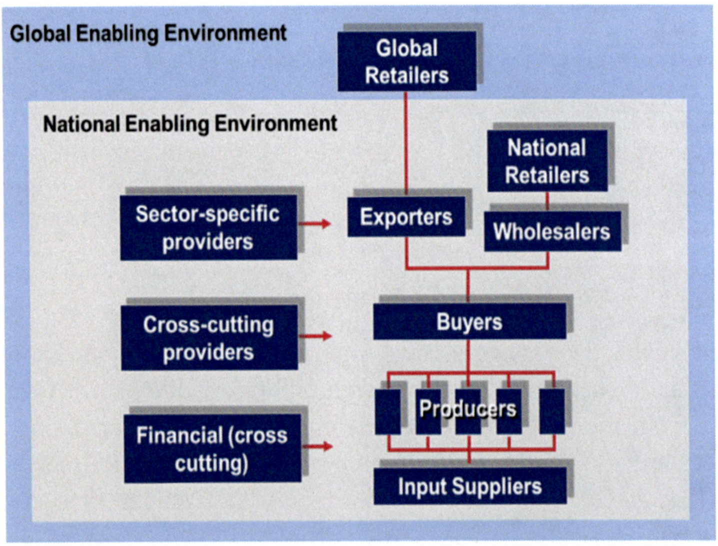

Figure 1: General Value chain structure.

Source: Dunn, (2014).

- **End markets:** End markets are always the beginning point for value chain analysis since the end market's demand determines the value chain's opportunities. As seen in the value chain map, a value chain can have many market channels, such as one serving domestic consumers and another serving consumers in export markets.
- **Vertical linkages:** Vertical linkages link businesses at all stages of the value chain, from input suppliers through manufacturers, processors, wholesalers, distributors, exporters, and so on, all the

way to retail. The commercial relationships involved in moving a product up the value chain are known as vertical linkages.

- **Horizontal linkages:** Firms at the same level of the value chain are linked together by horizontal links. Producer groups and exporter associations are two instances of horizontal links. Product aggregation to minimize transaction costs, more cost-effective access to inputs and services, access to information, and the empowerment of small businesses and farmers to advocate for change are all important functions of horizontal linkages at the producer level.
- **Supporting markets:** Firms and organizations that provide business support services to the value chain's firms make up the supporting markets. Many services, such as banking, transportation, and communication, are cross-cutting in that they serve companies across multiple value chains. Other supporting markets, such as enterprises providing technical advice and specialized services, are sector-specific.
- **Business enabling environment:** The formal and informal norms that help define the context in which decision makers work make up the business enabling environment. International food safety standards, national trade policy, inflation rates, natural disasters, municipal restrictions, and cultural and social norms are all examples of significant impacts in the enabling environment. Local, regional, national, and/or worldwide business enabling environments can all be identified.
- The emergent qualities of the market system correlate to the dynamic features in the value chain framework. Value chain governance, partnerships between enterprises, and the pace of learning and upgrading are some of the dynamic elements with crucial consequences for project implementation.
- **Value chain governance:** Power and the ability to exercise control over the conduct of other actors in the system are referred to as value chain governance. A "lead firm" is a company that sets, monitors, and enforces the rules that other companies follow. Information of the system's governance structure can help identify potential actions and establish routes for the development, transfer, and distribution of knowledge that leads to innovation.

- **Inter-firm relationships:** Interfirm connections that promote healthy competition and value chain competitiveness are mutually advantageous, collaborative, and allow for the free flow of information and upgrade incentives. Inter-firm interactions play a significant role in the transmission of knowledge and incentives, which improves firms' responsiveness to changes in end-market requirements.
- **Upgrading at the firm level:** At the firm level, upgrading is defined as investments that increase value contributed. A produce distributor, for example, might invest in chilling equipment to extend the product's shelf life. In order to develop a new variety of tomatoes that fetches a better market price, a farmer might invest in additional inputs, training, or irrigation. Upgrading can be done in a variety of ways, including:
- **Process:** greater efficiency
- **Product:** higher product quality
- **Function:** new role in value chain
- **Channel:** new market channel
- **Chain:** new value chain

In summary, a value chain approach adopts a market system view, focusing on the critical role of end market demand and identifying leverage points where interventions might catalyze increased value chain competitiveness. Interventions use a facilitation strategy to encourage value chain players to modify their behavior, form new or more cooperative relationships within the value chain, and/or improve the business enabling environment. Firm-level upgrading, value chain competitiveness, and a more broadly distributed distribution of benefits are all goals of interventions. Take, for example, Guatemalan high-value agricultural goods such as broccoli, snow peas, and baby veggies to show the structural and dynamic elements of a value chain. These labor-intensive crops are farmed with home and hired labor in the mild climate of the western highlands. There are several processing plants and exporters, some of which date back to the 1980s.

The value chain depicted in Figure 1 emphasizes some of the market system's structural characteristics. This value chain has several market routes, with end markets reaching customers both inside and outside of Guatemala, as shown in the diagram. Smallholders and microenterprises dominate value chain tasks such as farming, intermediates, and market vendors, as depicted in the darker portions of the picture.

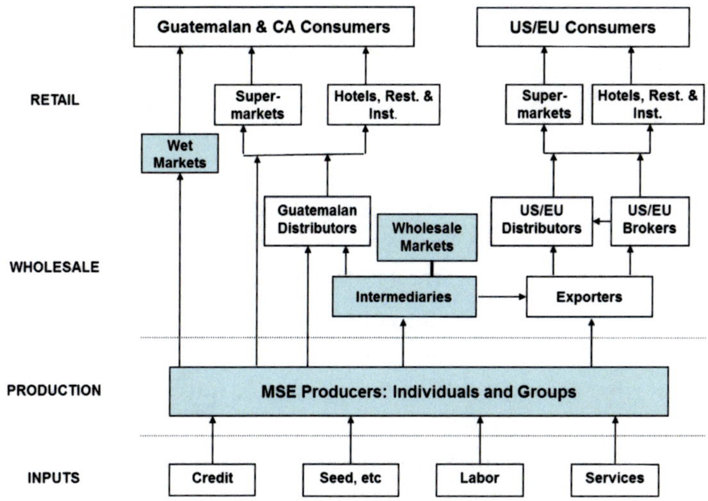

Figure 2: Value Chain Map for Guatemalan High-Value Horticulture.

Source: Dunn, (2014).

Smallholder producer groups created to transport goods to market, acquire inputs in bulk, or have better access to training and information are examples of horizontal linkages. The value chain map also depicts vertical linkages between various sorts of businesses, such as those between wholesalers and supermarkets in both domestic and international markets. Supporting markets are depicted at the same level as input providers at the bottom of the map in this scenario. The functional levels of the value chain, in this example inputs, production, wholesale, and retail, are located on the left side. In the case of Guatemalan horticulture, smallholder upgrading is focused on farmers meeting food safety regulations for fresh product entering the United States and the European Union. Food safety and product quality regulations in export markets play a major role in developing value chain governance in other horticultural scenarios. Additional governance rules derive from the need for packers and processors to schedule dependable high-quality product suppliers in order to meet their own delivery obligations. Buyers may find it in their best interests to provide embedded services that aid smallholders in adopting modern production and post-harvest techniques since they are incentivized to maintain a consistent stream of high-quality products.

VALUE CHAIN MODELS

Farm - Processor/Retailer Contracting (VC Initiated by Downstream Buyers):

Vertical coordination by downstream purchasers to farmers is the most prevalent value chain paradigm. This coordination can range from weak trading relationships between buyers and preferred suppliers to marketing contracts involving transaction agreements or production contracts requiring greater coordination. Production contracts may include specifics about the manufacturing process as well as the product's quality features. Technical and management assistance, as well as inputs on credit or cash credit, are sometimes included in production contracts to help contracted farmers satisfy quality, safety, and quantity criteria. In the traditional development literature, such interconnected contracts for input provisions are well recognized (e.g., Bardhan 1989). The logic behind such schemes is that a downstream corporation may have greater credit access than a farm because it has more collateral or cash flow to finance the investment and reduced transaction costs. The latter is possible when the lead firm invests in many suppliers (for example, as part of an out-grower strategy) and reaps the benefits of economies of scale. Another reason a downstream firm might be able to help its upstream suppliers is that they are closer to the final consumer and thus have a better understanding of consumer preferences in terms of quality and sustainability. The contract usually includes a commitment to adhere to buyer criteria (which could be public or private) as well as a transfer of inputs (or credit, to allow for quality upgrades) related to a purchasing agreement. Payment for these services is often made when the product is delivered. Simple inputs like as specific seeds, fertilizer, or animal feed can be provided. However, significantly more complicated types of knowledge transfer have been observed, particularly in fields where product quality is becoming increasingly crucial and long-term investments are necessary. Long-term technological advancements can be provided through extension services, technical and management help, quality control, specialized transport and storage services, investment loans, and investment assistance programs in more advanced types of contract farming. In several industries, buyer-initiated VC and contract-farming are popular. Contracting is usually initiated by milk processing firms in the dairy industry. They establish milk collection stations and milk delivery partnerships with farmers (individual farmers or farmer cooperatives). These contracts might be flexible or include more stringent selling and buying requirements. The desired quality in terms of fat and protein content, as well as the acidity of the milk, is frequently stipulated in milk contracts. Contracting between supermarkets, processing businesses, or exporters on the one hand, and farmers on the other is also widespread in the fruit and vegetable sectors.

There are numerous examples of the value of interconnected contracts between farmers and merchants, exporters, or processors. In order to provide cows to their contracted farmers and grow their operations, certain milk processing businesses in China (Sanlu Group Co. & Jinniu Group Co.) created their own breeding programs or integrated breeding farms in the VC system. They also help farmers obtain bank loans in order to purchase cows, and they act as a guarantee for the loan. They provide technical help, veterinary services, and feed to farmers. As a result, multi-stakeholder VC models with high levels of vertical coordination become highly complex. The provision of inputs (such as seeds and specific fertilizer) as well as extensive systems of technical guidance and extension services to contracted farmers are documented in studies on horticulture export chains in Africa (Henson, Masakure, and Boselie 2005). According to Minten et al. (2009), poor farmers' access to technology inputs was a crucial factor in their decision to sign contracts with horticultural export enterprises. Many other African value chains, however, show similar interconnected contracting. Exporters and processors of cotton, rice, barley, and tobacco, for example, frequently offer seeds, insecticides, and fertilizer to their suppliers, according to Bellemare (2012). Several researches on Eastern Europe and Central Asia show extensive and elaborate value chain contracting systems in numerous industries such as sugar, dairy, barley, cotton, and others in the 1990s and 2000s. Cotton gins in Kazakhstan, for example, supplied cotton farms with not only seeds and fertilizer, but also water, with water irrigation systems serving as a critical technological input for farms (Sadler 2006). Dries et al. (2009) and Van Berkum (2007) present research on dairy contracting systems from a variety of nations, demonstrating considerable input and technology transfers. Credit, animal feed, and technical guidance are all important components, as are investment loans for better dairy cows and milk cooling tanks. Dries and Swinnen (2009) indicate that interconnected contracting had a significant impact on milk quality for both small and large farms in Poland. In Uganda's dairy industry, processors like Pearl Dairy supply farmers with high-quality storage equipment as well as instruction on hygienic standards, disease control, and breeding. The barley malt-brewing value chains, as documented by Van Herck et al. (2012) for Eastern Europe in the 1990s and relevant for SSA with investments by all the major international brewing companies in countries like Ethiopia, are another interesting multi-stage example of input transfers in value chains. To brew high-quality beer, they all face the challenge of obtaining sufficient high-quality barley and malt. 4 To improve malt quality, the entire value chain had to be upgraded. Brewing businesses devised transfer schemes including malting processors, barley farmers, and seed companies for this purpose. Seed supply and selection schemes, investment aid, and advise on post-harvest storage and treatment are all examples of farm help. These programs were successful in terms of enhancing both quality and productivity.

Models Involving Intermediaries

Intermediaries in the chain can also commence vertical coordination. Wholesalers can become dedicated or specialized by specializing in selling to a single sort of buyer, or even a handful of distinct companies. Vertical coordination is carried out by these dedicated or specialized wholesalers both downstream and upstream in the supply chain. They enter into agreements with supermarkets or exporters to offer produce that meets the buyer's quality requirements, and they typically have long-term relationships with farmers to deliver the requisite volumes and quality. Dolan and Humphrey (2000), for example, discuss the critical importance of specialized exporters in Kenya and Zimbabwe's horticulture export value chains. On the one hand, these exporters are required to work with importers to produce new items for stores and to improve packaging, storage, and transportation. They must, on the other hand, verify that fruit sourced from contracted out growers meets high retail criteria in terms of supply timing and reliability, product and production qualities, and food safety. Another option is concessionaires, who are hired by supermarkets to stock and manage specific products on supermarket shelves. Concessionaires are paid a percentage of the sales value of the produce under concession by supermarkets, and they are responsible for the supply and management of the product (e.g., slicing, packing, labeling). These concessionaires rely on preferred supplier systems upstream in the supply chain. More complicated forms of contracting, where input companies were part of an institutional architecture including numerous partners, were among the other value chain innovations where input companies were included.

Bringing multiple VC agents together: Triangular Structures

Many processors and merchants are hesitant to lend to farmers for large-scale, high-quality investments. The reasons are self-evident: they necessitate large sums of money, and as the quantity of outstanding debts grows, so does the risk of late re-payment or default. As a result, processing businesses have approached financial institutions to see if they can work together to provide loans to farmers so that they can make the necessary investments to fulfill the standard. Institutional designs and cooperation are referred to as triangle structures. If a financial institution lends money to a farm that has a supplier contract with the downstream company, the downstream company usually gives a guarantee to the financial institution. The guarantee is essentially a promise by the purchase company to assume the supplier's debt obligation in the event of default. The underwriting is limited to contracting farms and is for loans for quality improvement that are tied to the contract. Because the funding (loans) is now (at least partially) covered by the financial institution, triangular structures need a lesser financial commitment from the processor or retailer. The guarantee is also likely to cut the farmer's interest rate because it reduces the financial institution's risk. In a "triangular structure," the third party might also be an input provider. In this situation, the buyer gives the entity that sells the inputs a payment guarantee. The argument is pretty like the financial institution's scenario. Sugar processors in Slovakia (Gow, Streeter, and Swinnen 2000), retailers in Croatia for fruit and vegetable supplier investments in greenhouses and irrigation (Reardon, Vrabec, et al. 2003), and dairy processors in several East European countries have all implemented guarantee programs within triangular contracting structures (Dries et al. 2009). These examples have been thoroughly recorded because their effects on farm-level investment and productivity development were profound. We are also seeing the growth of such triangular systems across Africa. In Ghana, for example, the processor/exporter Profound Integration has committed to acquire fresh pineapples from six farmer cooperatives, while input merchants have agreed to provide those cooperatives with the necessary inputs on credit (Kolavalli, Mensah-Bonsu, and Zaman 2015). The costs of these inputs are directly covered by Profound Integration, who deducts them from the cooperative's payout.

Bringing multiple VC agents together: Special Purpose Vehicles

The usage of so-called "special purpose vehicles" is an even more complicated method of boosting value chain sustainability (SPVs). A SPV is a separate corporation owned by many parties, such as a processor, an input source, and a bank. The SPV will usually contract with the farms after that. Output, input, and credit provisions can all be included in the contract. The partners in the SPV now share the risk of contract breach, which is a significant benefit of such an institutional arrangement. Although both the input provider and the financial institution gain from these contract innovations, when a downstream company implements input provision programs on its own, it bears the entire risk of farmers breaching contracts. SPVs, for example, allow for the spreading of risk among several agents, which encourages investment by corporations that might otherwise be put off by the danger. The collaboration between the Russian dairy processor Wimm Bill Dann (WBD) and the Swedish dairy equipment seller De Laval is an example reported in the literature. The purpose of the collaborative initiative "Milk Rivers" was to improve the quality of Russian farm products. They formed a joint venture, or SPV, to lease combine harvesters, milking, and cooling equipment (World Bank 2011). Serova and Karlova (2010) discovered that competitors of WBD began adopting the program to boost dairy farm investments a few years after the WBD-DeLaval collaboration got off. They employed a similar model (also with DeLaval), in which farms received milking equipment on a leasing basis, as well as calf milk replacers (CMRs) and feed additives, over a one- to five-year period. Horizontal spillovers have been documented in a number of examples (e.g., Gow et al. 2000), and they have the potential to disrupt the entire agriculture industry.

Vertical Integration (Complete)

In some situations, businesses have gone so far as to take over farming operations, i.e., "vertically integrating" their raw material supply. There are various reasons for this. One is high market exchange transaction costs or the danger of contracting holdups (Williamson, 1985). Increasing quality and sustainability standards, particularly in the private sector, may raise these transaction costs, especially if monitoring is expensive (e.g., restrictions on the use of pesticides and child labor). When farmers' capability is low and standards are complex (as is often the case with private standards), the expenses of input transfer and monitoring are increased (Gereffi, Humphrey, and Sturgeon 2008). Several studies illustrate how the growth of standards in high-value chains, as well as the resulting demand for farmers to invest in quality improvement, has resulted in the development of vertically integrated production systems. This has been documented in several research for Africa. In the Senegalese horticulture sector, for example, Maertens et al. (2011) show how a combination of available land and tightening public and private standards (such as HACCP and GlobalGAP) prompted exporters to shift from smallholder contracting to integrated estate production. In other parts of Africa, such as Ghana (Suzuki, Jarvis, and Sexton 2011), Zimbabwe (Henson, Masakure, and Boselie 2005), and Kenya (Henson, Masakure, and Boselie 2005), similar trends to vertical integration and big estate sourcing have been noted (Dolan and Humphrey 2000). It's worth noting, however, that in almost every case, the change to vertical integration was only partial, since processing companies continued to use a variety of sourcing routes. This method is motivated by several factors. For starters, land acquisition may be challenging due to practical (e.g., high population and farm density in fruitful areas) or legal limits (e.g., foreign ownership of land not being allowed). Second, social pressures (e.g., from local communities or worldwide civil society) could result in significant reputational consequences as a result of being linked to "land grabbing." Third, having a varied range of suppliers is an important aspect of a risk management strategy (Swinnen 2007). According to Suzuki et al. (2011), Ghanaian pineapple exporters combine own-estate production with smallholder sourcing to anticipate unforeseen market variations. Interestingly, substantial vertical integration has emerged in the large grain-producing countries of the former Soviet Union (Kazakhstan, Russia, and Ukraine), not so much to solve product standards as to overcome farmers' financial and input market constraints (Gataulina et al. 2006). Thousands of hectares are sometimes used by large agro-holdings. It's unclear to what extent they can serve as a model for large-scale investment in SSA's land-rich areas. In various Chinese fruit and vegetable export industries, export corporations and packing houses are increasingly opting for procurement from their own vertically integrated farms set up on government-leased property. According to Miyata et al., (2009), apple exporting companies source between 5% and 20% of primary produce from their own vertically integrated farms, while onions exporting enterprises source between 30% and 70%. Large-scale vertically integrated pig farms are expanding in peri-urban areas in various Asian nations, including China, Vietnam, and the Philippines. These businesses combine the various stages and operations of the supply chain into a single entity. Pig breeding, fattening, slaughtering, and meat processing are frequently combined in a single enterprise. These are frequently very large corporations that rely on imported genetic material and technology and are financed by foreign direct investment. In Uganda's dairy industry, different types of vertical integration can be seen. For many years, the processor Jesa Farm Dairy Limited has sourced half of its processing capacity (80.000 liter per day) from their own dairy farm to ensure quality and supply. The remainder is primarily derived from contractual dairy farmers to whom they pay premiums. While Jesa and the other cases mentioned above are examples of vertical integration started by a major downstream buyer of product, vertical integration can also start at the farm level. The Uganda Crane Creameries Cooperative Union, which represents over 100 primary dairy cooperatives in South-West Uganda, is constructing a processing facility in Mbarara. The required equity investment is funded by donations from members.

HOUSEHOLD ECONOMY AND SMALLHOLDERS

Smallholders are people who reside in rural parts of developing countries and farm a small piece of agricultural land. The size of a farm that is considered "small" is determined by the quality of local agricultural resources as well as the economic situation. Smallholder farms are those that operate on less than two hectares (Csaki & de Haan, 2003). In developing nations, there are over 2.5 billion people living on 500 million smallholder farms, with the majority living on less than $2 per day (IFAD 2013). According to data from the Food and Agriculture Organization of the United Nations (FAO), Asia has the biggest number of small farms (87%) and Africa has the second-largest number (12%). In Africa, there are around 33 million smallholder farms, accounting for 80 percent of all farms (Nagayets 2005).

Farm Size and Productivity

Smallholder farms have historically exhibited a productivity edge over bigger farms when assessed in terms of output per hectare, despite their inherent land constraints (Feder 1985). Smallholders have lower transaction costs in monitoring agricultural labor, according to the most widely accepted explanation for the inverse link between farm size and production. In many types of farming systems, smallholder households rely primarily on family labor and management, which has a competitive advantage over hired labor and management. This benefit is particularly noticeable in cropping systems that demand a lot of manual effort during the production process, such as horticulture crops. When cropping methods are more easily mechanized or routined, the benefits of family work in terms of oversight are diminished. The comparative advantage of smallholders in labor-intensive agricultural systems lays the groundwork for assessing their potential as producers in competitive value chains. Smallholder participation in value chains is also limited due to resource constraints in the wider farm family economy.

Farm Household Economy

One of the defining characteristics of the smallholder family, according to experts, is the interconnection of the household's production, consumption, and investment activities under a unified decision-making framework (Singh, Squire and Strauss 1986). Smallholders are agricultural producers as well as consumers, and, like other microentrepreneurial households, they make production and investment decisions in the context of their total household economic portfolio (Chen and Dunn 1996). As a result, smallholders' decisions to upgrade and engage in certain marketplaces are part of a larger process of allocating a household's limited resources among various production, consumption, and investment activities. This decision-making process can be treated as a restricted optimization problem with numerous objectives, such as income production, risk management, food security, immediate consumption, and long-term investment goals for school, housing, health, and key life events. Resource Constraints for Small-Scale Farmers Smallholders are not a homogeneous group; their resources and capacities vary greatly. Smallholders' resource levels and their ability to adapt to new market possibilities are linked by the family economic portfolio. Those with severe resource constraints may be unable to achieve the minimum participation standards. Land fragmentation, household labor shortages, and food insecurity are all examples of resource constraints that can impede smallholders' capacity to connect with more lucrative markets.

- **Land fragmentation:** Some smallholder households have such little land holdings that they can no longer be considered viable farming units. An examination of national surveys from five countries in Sub-Saharan Africa found that smallholders' farm sizes differed significantly (Jayne et al. 2010). Families with the largest landholdings also had the highest per capita income and agricultural earnings. Those with the smallest landholdings had the lowest crop earnings and were more reliant on non-farm sources of income. With fewer than 0.11 hectares of land per capita, at least 25% of households were on the verge of being landless.
- **Labor shortages:** When households are faced with chronic and/or cyclical labor shortages, their ability to adapt to new possibilities is hampered. During planting and harvest seasons, many smallholders face cyclical labor shortages. Hired labor enhances enterprise net profits for some smallholders while also adding to

rural employment. Chronic shortages of household labor limit the scope of the agriculture enterprise for other households. Due to a lack of domestic labor resources, women who lead smallholder households are frequently at a disadvantage.

- **Food insecurity:** Food hungry households are less likely to make hazardous investments, such as those required for upgrading. The requirement to improve food security can take precedence over farm enterprise investments for food poor households. A good example is when a family's food security concerns lead them to cultivate food crops when they could earn more money by raising higher-value commodities for market sale.
- **Financial constraints:** Even if households have enough food, they may be unable to participate in agricultural upgrading due to a lack of operating capital and liquidity for longer-term investments. New crops may necessitate greater economic outlays at planting time, while new techniques may necessitate physical assets such as irrigation pumps or greenhouses. The capacity to upgrade hinges on existing options in neighboring financial markets for those households that are unable to self-finance these investments.

In conclusion, as smallholder farmers in inclusive value chains, those households with severe resource constraints are less likely to find a way out of poverty (Fowler and Brand 2011; Garloch 2012). Instead, they might be better suited looking for work outside the farm. Before smallholders can make the investments and adopt the new behaviors required for upgrading and engaging in competitive value chains, they must have a certain minimum level of resource combinations.

UPGRADING AS BEHAVIOR CHANGE

Smallholders decide whether to upgrade based on their evaluation of the risk-adjusted returns to upgrading, in the context of their other options, resources and capabilities, and access to information and learning opportunities (Dunn et al. 2006). It's helpful to distinguish between factors that are internal to the household economy and ones that are external to the household yet part of the market system when analyzing smallholders' decisions to enter new markets and upgrade their investments. Opportunities and capacities are the two broad categories of elements that influence smallholder behavior change:

- **Opportunities refer to the perceived characteristics of economic alternatives:** Smallholders are motivated to enter new markets and spend in upgrading by external ("pull") incentives. These incentives are mostly dependent on smallholders' judgments of the new opportunity's profit and risk profile, as well as a comparison of that profile to the next best available option. End-market demand generates opportunities, but if end markets are far away, smallholders can access these chances by collaborating with other value chain actors. Incentives are filtered through the value chain and the business enabling environment in this way. It's the same as a smallholder examining a business opportunity and asking, "Is it worth it?" Is it a better option than my next best option?"
- **Capabilities refer to the capacity of the household to respond to opportunities:** Capabilities are based on the features of the household's economic portfolio and relate to the household's knowledge, skills, and resources. These include physical resources like land, irrigation, equipment, and machinery, as well as financial resources like working capital, savings, and credit, and human resources like health and embodied human capital like literacy, numeracy, and business and sector-specific knowledge and skills. Knowledge, resources, skills, and agency all contribute to capability. It's the same as a smallholder looking at an opportunity and wondering, "Am I able to enter this market and/or upgrade?" Do I have the skills I'll need to be successful?"

Changes in opportunities or capabilities (or both) that offer incentives for enterprises to respond are known as triggers. Smallholders' conduct might be influenced by changes in their awareness, understanding, and/or perception of opportunities. Changes in prices, changes in the laws of the enabling environment, access to more efficient technology and business models, fresh information, and better risk management procedures are all examples of triggers that change opportunities. A trigger, in other words, is an actual or perceived change in capabilities and/or opportunities that causes the person to be more willing (change in opportunities) or capable (change in capabilities) to adopt new behaviors.

INCLUDING SMALL-SALE FARMERS IN THE VALUE CHAIN

Participating in contract farming with domestic processors, exporters, or overseas customers is a significant way for rural farm-households, particularly in developing countries, to gain from agri-food exports and enhanced value in the domestic and export sectors. The amount to which smallholder farmers are incorporated in contract-farming arrangements, as well as the influence that involvement in contract-farming has on their incomes and well-being, determines whether they partake in the benefits of trade. There is considerable concern that structural changes (such as tighter product and production standards) would result in smallholder farmers being excluded from contract farming schemes, and therefore from supplying value chains. Due to lower transaction costs in buying bigger quantities from fewer suppliers, contract-farming schemes may favor larger farms (Key and Runsten 1999). Smallholders may be excluded as a result of this shift toward vertical integration or sourcing from larger suppliers, which may be influenced by standards. Due to a lack of technical and financial capabilities, small farms may be unable to meet demanding requirements, causing dealers and processors to reduce (or stop) procurement from small suppliers. In addition, when sourcing from smallholders, transaction costs for monitoring compliance with standards may be very expensive. Standards, on the other hand, are tools for harmonizing product and process attributes across suppliers, and as such, they can help cut transaction costs when dealing with many small suppliers. Furthermore, well-defined contracts that incorporate agricultural extension and assistance programs can help small farmers overcome financial and technical barriers to reaching demanding criteria. In fact, high-standard contract farming, coupled with tight contract coordination and expanded farm aid programs, may be a viable option for small farmers who are unable to participate in high-value export production. Furthermore, businesses may choose to work with smaller farms because they may offer a cost advantage – particularly in labor-intensive industries with tiny economies of scale, such as fresh fruit and vegetable production – or because contract enforcement may be less expensive with small suppliers.

The empirical evidence is contradictory (Reardon et al. 2009). Several empirical studies have shown that when standards rise, the proportion of export produce sourced from small farms decreases. In Kenya (Dolan and Humphrey 2000) and Cote d'Ivoire (Dolan and Humphrey 2000), for example, research show that smallholders are less involved in food export chains (Minot and Ngigi 2004). In response to rising private standards, Subervie and Vagneron (2013) explain the emergence of big exporter-owned lychee plantations in Madagascar. With the rising prominence of private standards, particularly GlobalGAP, Maertens and Swinnen (2009) demonstrate a transition in the vegetable export business in Senegal from smallholder contract farming to vertically integrated farming on large-scale plantations. Furthermore, Schuster and Maertens (2013) conclude that the spread of private standards in the Peruvian asparagus export sector, particularly production standards such as GlobalGAP, has resulted in decreased sourcing from smallholders, with certified companies sourcing significantly less from smallholders than non-certified companies. Some export industries, such as the tomato export business in Senegal, are entirely reliant on vertically integrated agro industrial farming, with no smallholder suppliers (Maertens, Colen, and Swinnen 2011). Other studies, on the other hand, suggest that smallholders are nonetheless included, if not entirely, in current value chains. Several studies from Eastern Europe, for example, show that small farmers are part of modern agricultural value chains (e.g., Dries et al. 2009, 2009). Smallholders have also been successfully integrated into numerous value chains in Africa and Asia. According to Minten et al. (2009), Madagascar's vegetable export sector employs 10,000 smallholder farmers and is totally dependent on an intensive contract-farming system. Fruit and vegetable sectors in Zimbabwe (Henson, Masakure, and Boselie 2005), Chile (Handschuch, Wollni, and Villalobos 2013), and Thailand (Handschuch, Wollni, and Villalobos 2013) are other examples where smallholders are heavily involved in high-value export chains through contract farming with buyers and exporters (Kersting and Wollni 2012). Smallholders also play an essential role in these value chains throughout Asia. In China, for example, export horticulture chains are nearly entirely reliant on smallholder contract production (Wang et al. 2009). According to Gulati et al. (2007), smallholder producers dominate numerous value chains in Asian countries, owing to contract farming and innovative vertical coordination schemes. In conclusion, empirical evidence on the exclusion/inclusion of smallholder producers in global value chains through contract-farming schemes across industries and countries reveals a mixed picture. In order to explain the various patterns of smallholder inclusion, Vandemoortele et al. (2012) established a formal theoretical model of the emergence of the demand for high-quality, safe food, as well as an analysis of which small producers are most likely to be included. They illustrate that certain producer are included in the high-quality economy, while others are excluded, depending on the economy's initial production structure, transaction costs, and the potential of contracting between producers and processors. Smallholders are more likely to be excluded in a mixed production structure comprising both smallholder farms and larger farm operations, according to their model. When the farm sector is more homogeneous and dominated by small farms, it is more likely that high-value production will emerge more slowly but more broadly. These predictions/arguments are in line with the findings of Reardon et al. (2009), who, based on previous empirical data, conclude that smallholders are disproportionately disadvantaged when sourcing from large farms is an option. They claim that lowering certain transaction costs (for example, through infrastructure expenditures, producer groups, and third-party quality monitoring) can help small and inefficient manufacturers better integrate into high-value value chains.

IMPACTS OF SMALL-SCALE FARMER INCLUSION

Efficiency Premiums and Spillovers in Smallholder Contract-farming: The extent to which smallholder farmers' engagement in high-quality export production and trade contributes to rural income mobility and poverty alleviation is determined by whether and how much contracted suppliers benefit from it. Foreign investors, huge food firms, and developing nation elites have long been accused of capturing the benefits of high-standard agricultural trade (Dolan and Humphrey 2000). Vertical coordination mechanisms and consolidation at the buyer end of export chains are said to increase the bargaining power of large agro-industrial firms and food multinationals, shift decision-making authority away from farmers, and strengthen these companies' ability to extract rents from the chain at the expense of contracted smallholder suppliers (Warning and Key 2002). Several empirical research, however, discover evidence that contradicts these assumptions.

Farmers benefit greatly once they are integrated in contract systems and high-value export chains, they discover. There is a lot of evidence of small to medium-sized farmers improving their welfare as a result of participating in high-value chains (almost always involving some type of contract farming), for example, in fruits and vegetables (e.g., Andersson et al. 2015), tobacco (e.g., Briones 2015), coffee (e.g., Bolwig, Gibbon, and Jones 2009), aquaculture (e.g., Hansen and Trifkovi 2014), dairy (e. (e.g., Jones and Gibbon 2011). Participation in contract farming systems around staple commodities like potatoes and rice has also been shown to have beneficial welfare impacts (e.g., Setboonsarng, Leung, and Stefan 2008). Ton et al. (2016) conducted a meta-analysis on the welfare impacts of contract farming, which found that contract farming has boosted the welfare of participating farmers by 62 percent on average, based on 26 qualifying studies (including several of the articles referenced above) (with a 95 percent confidence interval between 40 percent and 88 percent). Participation in high-value contract systems has been shown to improve food security, nutritional adequacy, and health in some studies. Participation in contract farming programs in more than ten different crops in Madagascar reduces the duration of the hunger season by an average of 8 days, according to Bellemare and Novak (Bellemare and Novak 2016). According to Chege et al. (2015), engagement in supermarket value chains by Kenyan vegetable farmers is related with a considerable increase in calorie intake (by 19%), vitamin A intake (by 96%), iron intake (by 18%), and zinc intake (by 18%). Increased income and a larger amount of land planted in veggies explain the favorable nutrition effect. However, because of the increased chance of male control of income as a result of increasing commercialization of vegetable production, these good impacts are partially muted (i.e., not as strong as they may have been).

According to Asfaw et al. (2010), farmers' health improves as a result of the use of less hazardous pesticides and better pesticide management as required by GlobalGAP criteria. Swinnen and Vandeplas (2012) develop a theoretical model to explain why, in high value chains, customers may pay suppliers an additional "efficiency premium" despite having uneven negotiating power in the contract relationship. The need for higher-quality products necessitates customers assisting farmers in improving production quality, for example, by giving finance for inputs. This offers holdup opportunities for the farmer, who can elect to use the inputs but sell the high-value output to another customer without repaying the credit given to him by the original buyer, which is likely in many developing countries. To avoid this, buyers are required to offer attractive contract terms in order to protect their returns on investment, such as a price premium to the farmer. As a result, even though the buyer has complete bargaining power, substandard suppliers can benefit from the implementation of quality criteria in a weak contract enforcement situation. Xiang et al. (2012) model the effects of rising food standards on household welfare in a general equilibrium setting. Their simulation results suggest that an increase in global or domestic demand for high-quality food results in an increase in high-quality product output, as well as a reduction in poverty and inequality. However, the study emphasizes the necessity of considering that the growth and equitable benefits of high standards are influenced by a complex set of factors and mechanisms that are sometimes overlooked in empirical research.

Technology Transfer and Access to Inputs through Value Chains: Contract farming usually necessitates the use of technology and input. Transfers are an important part of the benefits for local development because local suppliers lack the necessary skills, know-how, technology, management, money, and inputs. Farm aid programs can help overcome restraints on domestic enterprises in low-income nations with limited access to capital and technology in many circumstances, allowing these value chains to function. Several empirical studies, such as Dries and Swinnen's, document these technology transfers and the resulting productivity increases (2010, 2004). These studies show that transferring technology (and management) via value chains boosts productivity at the farm level, both for the product and for other production activities. Minten et al. (2009), for example, discover that superior contract-farming technology and management techniques flow over to other crops, resulting in considerable improvements in rice output and further improving the food security position of rural households.

Benefits for the Poorest and for Women through Labor Markets: Poor households may gain from job effects, which is an important and sometimes disregarded factor in agri-food trade welfare assessments. High-quality trade generates new jobs in labor-intensive produce processing and handling, as well as on vertically integrated estate farms and large contracted farms. A transition from smallholder contract farming to vertically integrated estate farming necessitates a shift away from family labor and toward paid labor. As a result, agri-food trade may provide additional benefits through employment effects. Although there is little empirical evidence on this topic, some recent empirical studies have shown that the development of such high-value agro-industrial value chains creates significant employment, such as in the vegetable export sector in Senegal (Colen, Maertens, and Swinnen 2012) and the cut flower industry in Ethiopia (Colen, Maertens, and Swinnen 2012). (Mano et al. 2011). The Ethiopian horticulture business presently employs 180.000 individuals, with 85 percent of them being women, according to the horticulture export organization (EPHEA). According to Webber and mLabaste (2009), about 7000 smallholders in Kenya were active in fresh vegetable export, compared to 40.000-60.000 people working in the processing business or as farm labourers. Employment in agro-industrial production and exporting enterprises is widely available for the poor in Senegal's vegetable export sector, and this employment has a significant positive impact on household incomes and poverty alleviation

(Van den Broeck, Swinnen, and Maertens 2017). Workers' working circumstances may improve as a result of the higher standards. Working circumstances may improve as a result of ethical or fair-trade standards. Barrientos et al. (2003), for example, indicate that, in some circumstances, labor rules and codes of conduct can boost workers' well-being. Even food quality and safety requirements, on the other hand, may benefit workers. Standards may result in higher salaries through an efficiency premium offered to trained personnel in order to encourage them to stay at the same company by raising the need for employers to invest in training. Following enterprises' certification to private standards in the horticulture export sector in Senegal, Colen et al. (2012) found evidence of expanded employment periods and higher remuneration for workers. Furthermore, there appears to be a considerable demand for female labor in these export industries (Maertens and Swinnen 2012). This has indirect consequences, such as higher child schooling (Maertens and Verhofstadt 2013) and investment spillovers, in addition to direct effects (Maertens, 2009). Agri-food export sectors contribute greatly to female empowerment in rural communities by providing off-farm work options for women.

ASSOCIATION AND INCLUSION OF SMALL-SCALE FARMERS IN THE VALUE CHAIN

Many reports emphasize the importance of farmers' associations in integrating smallholders into value chains and improving their negotiating power. The first reason is that incorporating small farmers into value chains reduces transaction costs. It may be able to cut transaction costs for purchasing technology, satisfying buyer needs, farmer training, quality control, and so on by contracting with a group of farmers (e.g., a cooperative) rather than individual smallholders. A farmers' association could play a significant role in (joint) quality control. For a variety of reasons, quality control is essential. It's critical to ensure that the buyer's criteria are met. However, if quality control is solely at the discretion of the buyer, there may be incentives for delays in the form of opportunistic quality control behavior. Farm groups can serve as a watchdog for food safety and quality standards (lowering transaction costs for buyers) as well as providing assurances about the accuracy of quality control (protecting farmers) (Moustier et al. 2010). A second point is that farmer associations can help smallholders integrate into value chains by reducing human and financial capital restrictions. Farmers are generally trained in certain production (and storage) activities while producing high-value items. Farmers' associations are frequently believed to have a role to play through extension services and other means. Experts, on the other hand, have expressed concern about this issue, arguing that the product and process needs in these value chains are frequently particular and go beyond the basic type of training that farmers groups typically provide, rendering VCD ineffective. A similar point is that smallholders may have better access to funding through farmers' associations, allowing them to make necessary investments that they otherwise would not be able to make. Farmers' associations may also be linked to credit cooperatives, which could help farmers with their financial needs even more. Farmers' associations' third primary justification in value chains is to increase farmers' bargaining power. Farmers may be in a stronger position in the supply chain to negotiate better contract conditions (and better policies) through collective action. Farmers associations' efficiency and equality effects are frequently emphasized as vital for improving smallholders' access to value chains and improving contract terms, hence making value chains more inclusive and lowering poverty among smallholder farmers.

Even though these points have been made in several studies (and for a long time), there is surprisingly little proof that;

- farmers associations play an important role in modern value chains, and
- that they are an effective institution to make modern value chains work and
- that they enhance benefits of small farmers.

Cooperatives have been shown to be a valuable tool, but they do not appear to be a straightforward solution for improving the value chain's influence on poverty reduction, partly because they exclude the poorest. There is some evidence that farmer associations can help smallholders become more integrated into high-value chains (e.g., Moustier et al. 2010). Less is known about the involvement of farmer associations in the development of (innovative) contract schemes10 or the mechanisms through which they influence VCD (e.g., reduced transaction costs in sourcing, improved access to technology, joint quality control, collective savings and credit schemes, or active knowledge exchange). According to Moustier et al. (2010), supermarkets prefer direct purchasing from cooperatives in the Vietnamese vegetable sector since it lowers prices (by eliminating collectors and wholesalers) and ensures better quality. They say that cooperatives can guarantee greater quality since they market produce collectively (e.g., by using a common label), actively manage quality, and provide better access to the government's agricultural extension office's training resources. Several studies have shown that joining a cooperative boosts the price earned by farmers (Chagwiza, Muradian, and Ruben 2016). This could be owing to farmers' increased bargaining power through associations, but that isn't always the case. Price levels are likely to be influenced by the cooperative's bargaining strength, as well as the quality of the produce, the type of product, and the degree of transaction expenses (all of which could be modified by the cooperative). Farmers' associations' overall poverty impact is also determined by whether the poor join in organizations such as cooperatives, and how much cooperative involvement by poor farmers affects their total household income. Obviously, the two factors are intertwined: individuals who do not anticipate to benefit from a cooperative are less inclined to join one in the first place. 11 On the one hand, one could anticipate the impoverished to benefit more in relative terms from collective action because they have the most to gain. On the other hand, because of their modest size, they are likely to profit less in absolute terms from cooperation, and the fixed costs of membership may be rather substantial. According to several research, asset-poor households are less likely to join a cooperative (e.g., Fischer and Qaim 2012). Social capital, human capital, landholdings, geographic location, and financial access all appear to be important. Furthermore, minimum land holdings are sometimes an explicit entry-criterion for joining a cooperative, meaning that cooperatives are purposefully excluding the poor in certain circumstances (e.g., Ito et al. 2012). In conclusion, the empirical data imply that farmers groups can help smallholders become more integrated into modern value chains and reap more rewards. However, studies reveal that the poorest people may not join cooperatives and hence may not profit from their participation.

GOVERNMENTS, DONORS AND NGOS INITIATIVES IN DEVELOPING INCLUSIVE VALUE CHAINS

In developing nations, governments, donors, and non-governmental organizations are increasingly realizing the benefits of well-functioning and inclusive value chains for economic growth, agricultural production, job creation, and poverty reduction. As a result, developing nations are progressively incorporating inclusive value chain development (VCD) into their agriculture and development policies and initiatives. Selective interventions focused at specific value chains or individual players inside those value chains can be separated from general interventions that target the (business) environments in which value chains function. Infrastructure, property rights, contract enforcement, corruption, and the administrative barrier of doing business are among the untargeted initiatives. Investments in human resources, public agronomic research, and general extension services are also included. These kinds of treatments aren't always acknowledged as VCD therapies, and they're rarely advertised as such. They are, nevertheless, a crucial instrument for enhancing the performance of value chains since they can minimize transaction costs and eliminate bottlenecks. Furthermore, a favorable business climate may render selective VCD obsolete. There are numerous types of VCD that include directly assisting one or more value chain participants. Below, we examine and categorize the various techniques based on the initiative's major engagement point. A farmer, a downstream lead firm (processor, trader, or retailer), a service or input provider, or a platform, cluster, or committee comprising numerous stakeholders might all fall under this category. Initiatives might be tightly focused or have a broader and more integrated strategy, depending on the entrance point. The implementation agency (public, semi-public, or private), the financing modality (grant, subsidy, or (concessional) loan), and the level and length of public involvement are some of the other parameters on which initiatives differ (one-time or continuous).

Value chain development initiatives directed at farmers: Direct help to farmers by an NGO or public extension agency is one of the easiest means to establishing inclusive value chains. Due to their limited capacities, lack of access to money, technology, and output markets, and their small size, farmers are frequently the weakest link in the value chain. Furthermore, because many VCD efforts are driven by increasing the well-being of smallholders, they are an obvious place to start. VCD directed towards farmers, in its most basic form, entails training farmers on agricultural methods that are important for increasing productivity or complying with buyer criteria. This strategy is now frequently supplemented by initiatives that assist farmers in forming groups for collective trading in input and output marketplaces. The fundamental idea is that cooperation improves farmer bargaining power by increasing access to input and output markets. Many farmer field school projects, such as Danida's Agricultural Sector Programme Support in Bangladesh, use this strategy. During one crop season, the Department of Agricultural Extension's project delivers a variety of training modules on simple agronomic skills, homestead gardening, nutrition, and group formation to poor and small farmer households.

VCD that is more integrated and intense improves farmer links with other upstream and downstream value chain participants. By representing farmer groups in coordination and negotiation with other input distributors and possible customers, or by assisting important service providers, for example (e.g., in collection, storage, transport, distribution of inputs). The NGO Solidaridad, for example, uses this strategy in the horticulture, aquaculture, and dairy sectors in South West Bangladesh. They have a five-year intense program in which they coordinate their farmer aid with buyer demands and requirements. The initiative also provides technical and financial assistance to disadvantaged landless households in project villages who want to start small businesses offering services to local farmers. Input distribution (small shops), output collection, and transportation are all included. Solidaridad's collaboration with a prominent domestic retailer (Agora) to deliver a high-quality mango variety to the shelves of Dhaka supermarkets is a successful example of the organization's work. This mango comes from contracted farmers who are trained in integrated pest management and post-harvest procedures by the initiative (i.e., hygienic handling of produce, grading, sorting, and packaging). The store considers the pilot project a success since it shown that product differentiation for the home market may be a successful approach. Solidaridad anticipates further spillovers in the form of increased contract farming in high-value fruits and vegetables, as well as further product differentiation at the store level. The longevity of the created relationships between farmers and the rest of the value chain is a major problem for these programs. When the NGO's public financing runs out, it will be forced to quit defending these farmers. The farmers' ability to keep themselves organized and keep agreements with input providers and customers in place will next determine if the linkages can be perpetuated. This necessitates the development of an exit strategy for these types of initiatives in order to ensure their long-term success.

Value chain development initiatives directed at downstream lead companies: Finance is required for the development of a private-sector-led value chain. As a result, funding downstream lead enterprises to develop the value chain(s) from which they obtain their produce is an alternate option for donors and governments to assist value chain transformation (i.e., making them more inclusive and sustainable). These kinds of interventions have become increasingly common in recent years. IDH (the Sustainable Trade Initiative), which is supported by the Dutch government, is one example of a large-scale intervention focused on global value chains. IDH forms public-private partnerships with leading agri-food multinationals (e.g., Mars,

Unilever, Cargill, IKEA) around several high-value commodities (e.g., cocoa, tea, coffee, cotton, flowers, aquaculture, etc.) to introduce voluntary sector-wide sustainability standards, such as UTZ certified and Better Cotton (Omta et al., 2014). IDH advocates for sector-wide improvements in environmental effect, farmer income, and labor conditions at a pre-competitive level, but also financially supports firm ideas for value chain development that go beyond certification. For example, IDH has integrated Rainforest Alliance certification training into broader Farmer Field Schools in conjunction with Unilever and the Kenya Tea Development Agency. In addition, over 30 chocolate firms have applied for financial support for their field projects through the Cocoa Productivity and Quality Program. This initiative certifies and trains 164.000 cocoa farmers (mainly in Ghana and Ivory Coast) on agronomic and post-harvest procedures (Omta et al., 2014). Another example is the private sector window of the International Finance Corporation's (IFC) Global Agriculture and Food Security Program (GAFSP). IFC provides multiple concessional loans to processors through the GAFSP fund to help them start or expand processing facilities in poor nations. This frequently entails the creation or expansion of a high-value sourcing system, as well as aid to farmers in meeting high-value requirements. Working with huge lead companies ensures scale and reach. It does, however, come with several risks that must be addressed. First, state funding for private-sector VCD activities may not be necessary in addition to measures that would have been undertaken by the private sector on its own. As a result, "additionality" of public spending is a requirement for efficacy. Second, using public funds to fund private VCD could lead to a dependency relationship, especially if farmers are trained to follow company-specific standards. As previously stated, if requirements are company-specific, it will be more difficult for farmers to profitably side-sell their produce to other purchasers, reducing the incentive for the lead company to pay a price premium. Finally, several NGOs and multilateral organizations wonder if VCD projects conceived and implemented by major lead businesses serve the farmer's interests best in general. As a result of these concerns, some private sector development programs, such as the Facility for Sustainable Entrepreneurship and Food Security, now require the participation of a non-governmental organization (NGO) to represent the interests of farmers and wage laborers and to ensure the initiative's inclusiveness. However, it is questionable whether NGOs can perform this function, particularly if they are financially reliant on these partnerships and serve as service providers inside the scheme (e.g., providing farmer training).

Value chain development initiatives directed at service providers, intermediaries, or input distributors: VCD programs can sometimes only help a few important intermediaries, service providers, or input distributors along the value chain remove or lessen certain restrictions for farmers in order to improve their welfare. One example is Wageningen University's Integrated Seed Sector Development Initiative in Ethiopia, which aims to improve and expand smallholder access to high-quality, high-yielding seed. The initiative advocates for legislation that allows for private production and commercialization of improved seed varieties in addition to the government-controlled channel, and it is establishing local cooperatives to produce better varieties that are appropriate to local agro-climatic conditions. Working on one value chain link at a time, these projects are frequently the least integrated types of value chain development. If you simply lift one farmer restriction, you risk making other constraints bind instead. The impacts of better seed use on agricultural yield, for example, are influenced by complementary agronomic techniques, the agro-climatic setting, and other input and output market flaws. As a result, this form of intervention is only appropriate as part of a comprehensive strategy (as in the case of Solidaridad in Bangladesh) or in a favorable environment with a single fundamental binding constraint.

Multi-stakeholder platforms: Multi-stakeholder platforms (also known as "agri-business clusters" or "value chain committees") are an alternative to providing individual value chain actors with direct support. Instead, the overall goal of multi-stakeholder platforms for VCD is to improve coordination among value chain actors, identify common interests, share expertise, establish new economic prospects, and operate as an advocacy organization for its members' common interests. Value chain actors aren't the only ones who can join these platforms; governmental and non-governmental bodies, as well as knowledge institutes, can also be members. Devaux et al., for example, discuss this method (2009). They explain how multistakeholder forums in Bolivia and Peru's Andean potato sector, sponsored by the Swiss Agency for Development and Cooperation, resulted in various product improvements, which in turn sparked new value chain organization models to meet the new quality criteria. These activities, according to reports, strengthened collaborative action, teamwork, and creativity, resulting in higher farm gate prices, more farmer earnings, and more stable markets. They also cite horizontal spillover effects as a result of other value chain participants imitating the platforms' offerings. Furthermore, as a result of the program's success, policymakers and donors are boosting their support for future collaborative value chain innovation efforts. According to Kolavalli et al. (2015), a multi-stakeholder platform in Ghana's pineapple industry resulted in a triangle value chain structure in which a financing institution offers farmers with loans for inputs required to meet current processor standards. After deducting the cost of the inputs to repay the credit to the financial institution, this processor pays the farmers directly.

Lesson In Implementing Value Chain Inclusion Among Small-Scale Farmers

Various pieces of literature that examine the implementation of inclusive value chain development activities can teach us a few things.

- The requirement for quality improvement and/or guaranteed supplies typically drives value chain development. This was especially evident after the economic reforms in Eastern Europe, where demand for high-quality items outpaced supply due to abrupt and strong competitive forces and Western European FDI. Following the increase of high-value exports, urbanization, and a rise in domestic purchasing power, similar market dynamics are now occurring in Sub-Saharan Africa and other emerging sections of the world.
- To address supply issues and weak public institutions for controlling exchange, private contractual initiatives have evolved. Quality supplies are difficult to come by for food retailers, traders, and processors. The lack of state institutions necessary to facilitate market-based transactions, such as enforcing property rights and contracts, exacerbates the problems.
- Traders, agribusinesses, and food firms enter contracts with farmers in exchange for assured and high-quality supply. Conditions for product delivery, prompt payments, and farm support programs for suppliers are common features of successful vertical contracting. Input supply programs, investment aid, trade credit, bank loan guarantees, extension and management consultancy services, and other types of farm help are available.
- These technology transfer programs were put up amid difficult circumstances. Successful programs are built on deep understanding of the industry and local conditions, and they provide the ideal conditions for successful and self-enforcing contracting. Furthermore, these programs must be adaptable

enough to change contract conditions in response to changing circumstances, which is a common occurrence in developing and rising economies.

- Many institutional technology transfer innovations employ both a pull and a push tactic. The push strategy entails making technology more accessible. Various modalities are frequently utilized for various types of technologies. Interconnected contracting or payment guarantees in a triangle system involving technology providers are frequently used to offer intermediate inputs. Longer-term investments, such as machinery, frequently necessitated a more complex contract structure, such as a triangular guarantee structure with a financial institution that would provide a loan to the supplier to purchase the equipment, or special purpose vehicles in which both the buyer and the technology company collaborate. The downstream lead firm frequently provides direct knowledge, training, and help. One benefit of this direct contact is that the downstream lead firm may oversee the supplier's production process and so contribute to contract enforcement.
- The pull technique entails offering higher incentives for technology upgrade investments. The most obvious approach to accomplish this was to charge a greater price for higher quality. Other techniques included building confidence (for example, by making timely payments) and making relationship-specific investments by the downstream lead firm to boost the "private enforcement capital." This reduced the danger of suppliers investing in relationship-specific technology.
- It is critical for the technology transfer program's originator to have access to funding. In each of the situations examined, the company that pioneered the technology transfer innovation had financial support from FDI or had substantial financial resources. Because interconnected contracts, pre-financing, and guarantees necessitate big upfront investments or adequate collateral, this is the case. International organizations have the capacity to play a significant role in this situation.
- These programs can have a large influence because they can change the entire value chain towards a higher equilibrium, affecting all agents. Spillovers aren't just for vertical interactions; they can also happen horizontally. Competitors of firms that

initiate a technology transfer program may introduce similar contractual arrangements, either to stay in business (as farms would otherwise shift to supplying other companies) or because it is profitable for them to do so once they see the innovations succeed elsewhere – or both. Contractual convergence of this nature may extend beyond the sectors where the transfer initiative was begun. Similar contracts may be offered by other sectors competing for the same resources (e.g., land), or financial institutions may standardize the technique for other farms.

- The issue of enforcement is a significant one. When there are no public enforcement organizations, enforcing the law is difficult. In many transition countries, trust is often inadequate as a foundation for business dealings. Companies attempt to construct "self-enforcing contracts" by drafting contract provisions in such a way that no one has an incentive to break them. They also use "interlinking markets" to try to enforce contracts. The output market is where the credit transaction (loan and repayment) is enforced. However, there have been numerous instances where enforcement has failed. Even in successful cases, the contracts had to be fine-tuned or adjusted as circumstances changed. Creating the necessary conditions for effective and self-enforcing contracting necessitates a thorough understanding of the industry and the local environment.
- Farms confront a variety of constraints, which are reflected in contract forms. For example, guaranteed access to markets, capital, and inputs is a major contract motivator for farms in Africa, as input market constraints are important.
- Successful value chain development provides significant direct and indirect benefits. The processing company that launches vertical contracting will see an improvement in output and productivity. Contract assistance measures have a positive impact on farm productivity and product quality in the long run. Specialist storage (cooling equipment in dairy), veterinary support, and physical inputs had the biggest impact on output. Payments on time, fixed prices, and market access all had a significant beneficial impact. Specific programs resulted in a significant improvement in output quality. Farm investment was boosted by direct loans and loan guarantee programs. Indirectly, programs that aid farms in obtaining inputs (mostly feed) increase

investment by cutting input costs or minimizing transaction costs in obtaining inputs, hence enhancing profitability.

- Competition has a variety of implications on the development of value chains. It may help to distribute equitable and efficiency gains, but it also makes enforcement more difficult. In supply chains, competition is critical for equity and efficiency. First, competition causes VC spillover effects throughout the industry, as other processors are driven to implement comparable supplier assistance programs, since suppliers may refuse to supply unless they are offered identical terms. Second, competition keeps processing corporations and input suppliers from wielding monopoly power over farm contract terms.
- Successful contracting and supplier support programs are driven by foreign direct investment (FDI). It is a catalyst for institutional transformation and innovation. Because foreign corporations tend to pay more attention to quality standards, more sophisticated forms of vertical integration with a stronger emphasis on quality and standards are frequently introduced. Spillover effects, on the other hand, lead to convergence when domestic enterprises begin to mimic the management methods of overseas affiliates.
- Inclusion of small farmers. First, in supply networks, transaction costs benefit larger farms. Second, small farms face greater financial constraints when it comes to making critical investments. Third, small farms require more support per unit of output than larger ones. As a result, businesses seek to work with fewer, larger, and more contemporary suppliers.
- Businesses operate with a surprising number of suppliers, all of which are of a surprising modest size. There are numerous explanations for this. If small farmers account for most of the supply base, companies may have no alternative. With larger farms, contract enforcement may be more difficult. In farm-processor partnerships, farm willingness to learn and attitude are more essential than size. In labor-intensive production activities, small farms may have a cost advantage. Processors may prefer to work with a variety of vendors. Small farms are more inclined to work with cooperatives than large corporations, domestic or foreign.

- The Conundrum of Small-Scale Integration Small poor farms may fare best (in terms of "supply chain driven development") if they are situated in an environment dominated by small poor farmers. If small farmers rely on agricultural assistance packages to make necessary upgrades, there will be a problem if there are insufficient (quality) supplies because the processor is unlikely to come up with VC packages.
- Poverty alleviation by the establishment of jobs on larger farms. The public policy discussion presents the issue of vertical integration and small farms (explicitly or implicitly) in terms of how public policy may prevent large, sometimes international, agribusinesses from exploiting (small) farmers in their contractual agreements. Reality, on the other hand, paints a considerably more complex picture. First, while all agribusinesses tend to prioritize profitability, this does not appear to lead to farm exploitation. There is more evidence that producers gain from VC than that they are abused in circumstances where there is enough competition.

Chapter 3

Profitable Small Scale Farming Ideas

CONTENTS

Introduction....72
Microgreens....73
Mushrooms....89
Soft Fruit / Fruit Trees Small Gardens....102
Micro Dairies....115
Small-Scale Chickens & Eggs Farming....121
Small-Scale Beekeeping....144
Fish Farming....154
Snail Farming....164

INTRODUCTION

Agriculture was a part of daily life in the oldest civilizations, and it provided food for communities and tiny groups. With a reliable food source, we were able to progress from hunting and gathering to a more refined and sophisticated way of life. Because of the growth of society from nomadic to one with permanent settlements and structures, other forms of livelihood became conceivable. When farmers worked the land to support the community, the community relied primarily on those farmers to provide a substantial portion of what they required, small-scale agriculture became the core of many communities. Small farms have recently been regarded as the backbone of the agricultural economy in the United States, with family farms accounting for 97 percent of all farms in the country. According to the USDA Economic Research Service, you must earn less than $350,000 per year to be designated a small farm. In the United States, there are about two million small farms, which comprise retirement farms, off-farm occupation farms, and farm-occupation farms. You can learn more about the USDA's small farm classification method on their website, but if your operation is between $0 and $350,000, you can be sure it will be classed as a small farm. Many modern-day farmers, particularly those who run big commercial farms in the Midwest, have passed down their knowledge and abilities from generation to generation. For tiny beginning farms, this isn't the case. These farmers had to learn how to farm by either apprenticing with other farmers, attending farm school, or putting in a lot of self-directed study time.

A productive farm does not necessitate a large quantity of land. Micro farmers are increasingly looking for property in urban areas that lack open space, but converting a small plot of land into a profitable enterprise requires more than a green thumb and a passing interest in gardening. Farming technology is continuously evolving, and staying on top of the latest methods and equipment is critical if you want to succeed in a competitive industry. Successful micro farmers treat their land as a business venture, embracing cutting-edge farming equipment to increase production and profits. Micro farms are small-scale agricultural activities that require significantly less land than a conventional commercial or family farm — often less than five acres — and are frequently found in urban or suburban areas. Micro farmers are forced to be innovative about the crops they raise and the way they seek to earn a profit because of size and zoning restrictions, and they tend to focus on sustainability, seasonal crops, and specialized markets for their products. Micro farms, despite their small size, can generate a large amount of food per acre, which, when correctly managed, can result in large profits. The ordinary maize or soybean farm, for example, earns $400-600 per acre, whereas a new wave of micro farmers led by Quebec's J. M. Fortier aims to earn $100,000 per acre. Modern micro farmers combine high-tech innovations with traditional practices in order to preserve the land's natural fertility. However, when it comes to profit yield, size still matters for some crops. To compensate for the lack of space, many micro farmers aim on maximizing their time, which means profitable growers prefer short-season crops that can be re-sown multiple times in the same piece of land throughout the season. This form of cultivation is exemplified by products like "baby" spinach and arugula. A conventional spinach crop can take up to six weeks to mature, however baby spinach leaves can be picked in as little as three to four weeks. Micro farmers can maximize their small plots for numerous rounds of spinach throughout the growing season by planning succession plantings of baby spinach. To sustain fertility, such extensive use of the soil necessitates rigorous crop rotation, interplanting, and composting, yet one rule stays true: Micro farmers may bring more produce to market throughout the season with short-season crops. Micro farmers should also keep an eye on their local market's hunger for specialist crops in order to capitalize on seasonal favorites that can fetch high prices. Specialty crops such as garlic, herbs, and mushrooms are examples of low-land-use crops. Heirloom tomatoes are famous for their unmatched flavor and may bring in large profits, especially when farmers focus on growing hard-to-find types, even though they require a longer growing season.

MICROGREENS

Overview

Microgreens are vegetable greens collected immediately after the cotyledon leaves have grown (not to be confused with sprouts or shoots) (and possibly, with one set of true leaves). They're utilized as a dietary supplement, a cosmetic enhancement, and a flavor and texture enhancer. Microgreens can be used to add sweetness and spiciness to a variety of dishes. Microgreens are smaller than "baby greens" because they are eaten shortly after sprouting rather than after the plant has matured enough to produce numerous leaves. They're currently considered a specialty greens genre among upscale supermarkets, suitable for garnishing salads, soups, sandwiches, and platters. For added taste and nutrition, they can also be used as a main vegetable in some recipes. They're used as a garnish in many recipes, but they're also used as the main ingredient in others. Garlic pea shoots, pea shoots or micro cabbage in cabbage soup, or coleslaw made with radish microgreens instead of cabbage are just a few examples. Microgreens are becoming increasingly popular as a result of their powerful flavor and nutrients, and chefs and cooks are coming up with novel methods to use them. Young greens can be made from a variety of vegetables, herbs, or other plants. They are 1 to 3 inches (2.5 to 7.6 cm) in length, including the stem and leaves. During harvesting, the stem is chopped just above the soil level. Microgreens often have one pair of very small, partially developed true leaves and fully developed cotyledon leaves. Microgreens first appeared on restaurant menus in San Francisco in the 1980s. Microgreens have been grown in Southern California since the mid-1990s. Arugula, basil, beets, kale, cilantro, and a colorful variety of those called a "Rainbow Mix" were among the first types available. They are currently produced in most areas of the United States, with an expanding variety of cultivars, after spreading eastward from California. Today, the microgreens sector in the United States is made up of several seed firms and growers. A central stalk, cotyledon leaf or leaves, and often the first pair of extremely young true leaves make up microgreens. They come in a range of sizes, ranging from 1 to 1.5 inches (25 to 38 mm) in total length, depending on the type grown. When the plant reaches this size, it is no longer classified as a microgreen. Petite greens refer to the larger sizes. Fast-growing microgreens, such as many brassicas, have an average crop period of 10–14 days from seeding to harvest. Microgreens that grow slowly, like beets, chard, and many herbs, can take 16-25 days to reach harvestable size. There is no legal definition for baby greens or microgreens. The words "baby greens" and "microgreens" are marketing terms for different types of plants. Sprouts are germinated seeds that, depending on the species, are eaten as a whole plant (root, seed, and shoot). Almond, pumpkin, and peanut sprouts, for example, are said to have a better flavor when picked before they form roots. Due to their comparatively greater risk of microbial contamination compared to other greens, sprouts are legally classified and subject to specific rules in terms of manufacturing and marketing. Growers who want to sell sprouts should be aware of the hazards and measures outlined in the FDA's Guidance for Industry: Reducing Microbial Food Safety Hazards for Sprouted Seeds (FDA 1999).

Starting A Microgreens Business

Growing microgreens can be a profitable business idea if well planned and implemented. It's no longer a secret that microgreens are becoming increasingly popular. Although the demand for microgreens is increasing, the supply is still not keeping up. At present time, this is the main reason for the relatively high rates of microgreens. A microgreens business requires very little money and time to get started. You can get started for as little as $100 and the business will just take a few hours of your time. With just 10-20 trays, you may easily make $500 – $1000 every month. But it isn't that simple. Climate, location, seed selection, quality control, crop diseases, bad financial management, and other factors can make or break your business. Growing microgreens presents some problems, but they are not insurmountable if everything is organized properly and deliberately.

Consideration for Starting Microgreens Business

When beginning a new business, it's always a good idea to plan, and anyone considering a microgreens business should take that advice to heart. It is significantly easier to establish the sorts and quantities of microgreens to cultivate, as well as how to market them, with proper planning.

- **Explore the Local Market:** Selling microgreens, like any other business, does not arise by accident. If you're new to microgreen gardening, it's critical to figure out whether area restaurants, co-ops, and grocery stores, as well as farmers market patrons, are interested in buying microgreens. Farmers at Microgreens have realized that local restaurants are ideal customers for their crops. Microgreens are utilized in a variety of ways, including garnishes, salads, and the creation of novel foods. Chefs all around the world are concerned about presentation, and microgreens are a terrific method to boost the appeal of menu items. Fresh, high-quality product is always in demand by food cooperatives. Health-conscious buyers who make it a point to use the best products available are quick to scoop up microgreens grown locally. Even great if the microgreens are organic. Local grocery stores, too, seek to obtain products from the area whenever possible. That means that anyone who can consistently deliver high-quality products will almost always find a willing consumer. If you're not sure if a local market will be interested in your microgreens, simply inquire with the produce manager about the possibility of selling them straight to the store. Finally, pay a visit to your local farmers market to see how busy it is and how many others are selling microgreens right now. Even if only a few people are selling them, if your market is vast and robust, there may be enough buyers to go around.

- **Pay Attention to Business:** All income and spending should be meticulously tracked by a successful microgreens business owner. Understanding fundamental accounting concepts and keeping track of every expense is the only way to build a truly profitable business. I truly enjoy the profit-first accounting system, which was popularized by the book of the same name. Most people who grow microgreens for sale start small to learn the growing process and evaluate if microgreen farming is something they'd be comfortable doing before moving on to selling and financial management. It's also simpler to start part-time and hold other employment until you're confident that the business will generate enough cash to sustain your lifestyle. While simple ledgers can be used to track income and expenses, it's usually easier to utilize a computerized bookkeeping system like QuickBooks or Wave. Modern digital bookkeeping enables the creation of business reports, the tracking of expenses and income, the preparation of taxes, and the evaluation of numerous metrics to determine which adjustments may result in increased revenue in the future.
- **Choose the Best Space Available:** A microgreens business, like any other sort of farming, necessitates a certain quantity of suitable space. While you won't need a lot of area or large-scale machinery, you will need the essentials. To cultivate great fruit, every microgreen farmer needs to supply light, water, and a temperature-controlled atmosphere. A simple greenhouse may give ample space and light for much of the year, depending on local weather conditions, but most microgreens producers will want to build an indoor grow room to have more control over the growth circumstances. To keep the development cycles continuing, you'll need a watering system, and we recommend installing a water line connected to a hose inside your grow chamber. Because microgreens are temperature sensitive, a heat source may be required in the winter to raise the temperature of the grow chamber to the mid-70s for warm-loving crops like sunflower and basil. We recommend a basic space heater with heat mats, such as this one, which can accommodate four 1020 trays. In the summer, keep your grow room below 80 degrees Fahrenheit and the humidity low, ideally in the 40-50 percent range. Mold and other diseases can be avoided by lowering

the humidity. A high-quality dehumidifier, such as this one, is recommended for this.

- **Tailor Your Crop to Meet Local Demands:** We've already discussed how to find some of the best local clients, but there's another approach to improve your microgreens company' chances of success. That example, rather than trying to cultivate a range of popular microgreens, the crops will be tailored to satisfy the needs of a particular market. Exotic cultivars such as Dark Opal basil, Bulls Blood Beets, and Red Garnet Amaranth may be requested by local chefs to give a splash of color to dishes. Doing some study is the greatest method to figure out what needs are present. Get out there and speak with chefs, finding out what they're looking for and how you can help. Visit local restaurants to find out what sorts of microgreens are readily accessible and which are difficult to come by. The key is to concentrate on the services that other providers do not currently provide. Because microgreens have such short growing cycles, starting seeds that will produce a more profitable harvest immediately is usually not difficult. However, keep in mind that some of these more exotic types might be far more difficult to grow. If you decide to take this path, be prepared for a lot of trial and error as well as crop failures. Also, expect to charge a premium for these high-end varietals. If you want to accelerate your learning, enroll in our Microgreens Business online course, which contains crop cheat-sheets for all microgreens kinds, even the more difficult exotic crops. Keep in mind that tastes and demands are continuously changing. As a result, a microgreen that is popular in the spring may not be so popular during the hotter summer months. While most people choose to cultivate the same microgreens, making the greatest money requires staying on top of current trends and offering fruit when it's needed by local customers.
- **Provide the Best Service:** Every consumer wants to know that the things they're buying are of the highest quality and that the provider is concerned about their demands. As a result, every successful microgreen farmer would continuously monitor the quality of their crops and ensure that orders are fulfilled on time. Restaurants, for example, may have weekly specials that call for more basil or mustard, or may only utilize greens on weekends.

They want the microgreens delivered as needed, so they're constantly fresh and appealing. It can be difficult to accommodate order modifications and add-ons on the fly, but as your business grows, you'll be able to grow a little more and have a buffer for last-minute alterations. Your microgreens purchasers care about price, but quality products will always win out over ones that have been harvested and stored for days. Customers can really save money by purchasing your more expensive goods because they last longer and produce less waste. One reason local grocery stores, restaurants, and food co-ops buy from local suppliers wherever possible is because of this. They will, however, swiftly distance themselves from microgreen growers that fail to offer high-quality crops. Daily, careful planning is equally as necessary as it is when launching a new microgreens firm. While there is some flexibility in the planting-to-table timeline, the top microgreen producers will always harvest most of their produce as close to the day it is intended to be consumed by the end customer as possible.

Benefits of Growing Microgreens as A Business

Here are some of the main reasons why I believe cultivating microgreens is an excellent business to start.

- **Low startup cost:** You can start with one or two 10″ x 20″ trays of microgreens and grow your business from there. In terms of soil and seed, each tray costs roughly $2. Start by supplying a single restaurant or producing enough microgreens to sell once a week at a farmer's market, then scale up as demand for your product rises.
- **Fast turnaround time:** Unlike corn or wheat, you don't have to wait an entire season or more to harvest these crops. Microgreens can be harvested in as little as one week and at most four weeks. That quick turnaround time allows you to test new ideas and improve your operation's efficiency. If your sales change throughout the year, you can quickly scale production up or down.
- **Year-round growing:** Microgreens are one of the few ways a small farmer can grow food year-round unless they have greenhouses. Another option is to grow mushrooms. This also means it's a reliable source of revenue. If you're already a farmer,

you may utilize microgreens to supplement your income and diversify your business over the winter.

- **Higher nutrition:** Microgreens are a type of superfood. They're chock-full of vitamins and minerals. They appeal to both restaurants and health-conscious consumers as garnishes and high-end ingredients.
- **High-value crop:** Top restaurants and grocery stores pay a premium for microgreens. As a local producer, you can charge a premium for them because they're such a unique product that's best when it's fresh.

Microgreen Productions

Microgreens require very little space to grow. If you have a spare bedroom in your home, you'll have plenty of room to get started. To make the most of their vertical area, most microgreen growers employ racks. Many growers can produce 50 pounds of microgreens in a two-week cycle using racks with four shelves in a 60 square foot facility. At $20 per pound, each cycle will cost $1,000, or nearly $2,000 each month. So, after your business is up and operating, you may earn $30 to $50 per square foot (about $100 to $160 per square meter) per month. To get to that stage, you'll need to invest in equipment, optimize your growing, and develop a regular client base. When growing microgreens on a single level, such as a tabletop, each 10″ x 20″ tray will normally yield 5–6 ounces (140–170 grams) of microgreens, depending on what you're producing. Microgreens manufacturing for household consumption does not necessitate the use of any specific equipment. The amount of time you'll need to commit to your microgreen business each week is entirely dependent on its size. If you're only cultivating a dozen flats every week in a spare room, it can be a part-time job that takes 15–30 minutes every couple of days, plus some time for harvesting and selling. Expect a full-time job and then some if you're running a six-figure (or more!) microgreens business. To keep up with demand, you'll almost certainly need to hire some new personnel. In other words, a microgreens business can take up as much or as little time as you desire. You get back a proportionate quantity of what you put in, like with most things in life. There are obviously some economies of scale that you may benefit from. For example, growing and selling 16 flats of microgreens takes no more time than growing and selling 8. Here are the bare minimum tools and resources you'll need to get your microgreens business off the ground.

- A 4-foot fluorescent light fixture with two bulbs. Along with standard T8 fluorescent bulbs that can be found in any store.
- 16 drain holes 10″x20X trays You can grow 8 flats of microgreens per week on a 4-foot setup. To add extra stability and strength, I recommend doubling up on your flats.
- Towels made of paper. When you're starting seeds, cover them with plastic wrap.
- A spray bottle, to be precise. To avoid drying out your seeds during germination and to keep your microgreens moist as they grow.
- Seed. I'd start with radish seeds because they're the easiest to cultivate and produce the quickest results. It's worth it to spend a little more money on organic seeds. You may even promote your microgreens as organic this way.
- A weighing scale. To prepare your microgreens for packaging, measure them out. A standard kitchen scale will suffice.
- A good pair of scissors or a large, sharp knife will suffice. Harvest your microgreens with this tool. A dull blade will make harvesting a headache, so it's worth spending a few more dollars here.
- Soil for pots. Any good organic potting soil will suffice.
- There's a clock. To program your lights to switch on and off at predetermined intervals.
- A tiny fan is used. To avoid mold and other problems, keep your crops well ventilated.
- A low-cost watering can. Choose one with a fine spout. You'll want to start watering your microgreens instead of misting them as they grow larger.

Your materials should only cost you a few hundred dollars in total. With an 8-flat setup, you may anticipate to make around $150 per week if you profit $20 every flat sold. I think 8 flats is a reasonable starting point. When you're just starting out, don't try to raise 20 varieties at once. Starting small and mastering one form of microgreen before moving on to the next is preferable. You'll definitely need more than 100 trays and 8 sets of lights, as well as a lot more seed, if you want to start a larger microgreen business. You'll most likely wind-up spending around $1,000. However, once you begin to sell your product, you may always reinvest revenues to develop your firm.

Steps to Grow Microgreens

Many different types of seeds can be used to grow microgreens. Seeds from the following plant families are used to make the most popular varieties:

- Brassicaceae family: Cauliflower, broccoli, cabbage, watercress, radish, and arugula
- Asteraceae family: Lettuce, endive, chicory, and radicchio
- Apiaceae family: Dill, carrot, fennel, and celery
- Amaryllidaceae family: Garlic, onion, leek
- Amaranthaceae family: Amaranth, quinoa swiss chard, beet, and spinach
- Cucurbitaceae family: Melon, cucumber, and squash

Microgreens can be made from grains including rice, oats, wheat, corn, and barley, as well as legumes like chickpeas, beans, and lentils. Microgreens have a wide range of flavors, ranging from mild to spicy, mildly sour to bitter, depending on the variety. Their flavor is generally described as robust and intense.

Step 1: **Calculate and measure the optimal number of seeds needed for your trays:** Following the instructions provided here or using the Microgreens Seed Density Calculator developed to make things very easy for you, calculate the number of seeds you need for your square, rectangular, or circular planting trays after selecting the species of microgreens you want to grow and purchasing microgreens seeds. You may measure the number of seeds you need using a tiny scale that you may already have in your kitchen. Once you've measured the number of seeds for a tray, you can use measuring cups or spoons to measure the number of seeds you need.

Step 2. Prepare your trays and growing media: Microgreens can be grown in trays and containers of various shapes and sizes. Flat trays are favored over standard nursery pots because microgreens do not require a lot of space to thrive. Microgreens planting trays of various sizes can be obtained from a variety of sources, however containers derived from food packaging can also be recycled. The major tip is to use material that is ideal for food production, is clean, and has drainage holes in the bottom, which will allow you to water the trays from the bottom without allowing the water to contact the greens while allowing the surplus water to escape. After filling or arranging the trays with the growing media, wet the growing medium from the bottom with a clean container without holes and filled with drinking water. Once the medium is wet enough, drain the surplus water. The wetness in the medium will aid in the adhesion of little seeds to the surface and keep them moist during the germination process.

Step 3. Seed your trays: You may now begin sowing by evenly dispersing the specified number of microgreens seeds across the full growing area of each tray. Most microgreen seeds do not need to be treated. However, if seeds are pre-soaked in water for some species with larger seeds or a strong seed shell, the germination process may be hastened. Seeds can be immersed in water for 8-12 hours overnight. It's a good idea to rinse the seeds in running water a couple of times during this procedure to wash them and give them some oxygen. It is not essential, and it may be preferable, to cover the seeds with dirt once they have been distributed on the growing media surface, in order to keep the sprouts clean. Following seeding, It's enough to spritz the seed with water every now and then to keep it moist during the germination process. It is recommended that the seeds be kept in a dark atmosphere for a few days to aid germination and maintain a good moisture level during germination. Covering the microgreens trays with something that can block the sunlight is a simple way to accomplish this. It is advisable to place some weight on the seeds throughout the germination process for species, such as pea and sunflower, to maintain the seeds in touch with the soil. Maintaining the weight on top of the shoots during the germination process aids the dissociation of the seed hulls from the cotyledons in sunflower and other species.

Step 4. Microgreen's growth and management: It is feasible to uncover the trays and expose the sprouts to sunlight once the germination process is complete and the seedlings have reached a particular height. When deciding where to grow your microgreens, keep in mind that, like any other plant, microgreens thrive from plenty of sunlight, so put them where there's plenty of it. A window on the south side of a house or apartment will often receive more sunlight than one on the north side. If the light is limited as the microgreens begin to grow, the shoots will bend toward the light. Plants create antioxidants predominantly in response to solar radiation, therefore a suitable amount of sunlight will ensure optimal growth and a higher buildup of antioxidants. Apart from the light, the only thing left to do now is make sure the seedlings have enough water. Water should only be used when essential, to avoid excess moisture and to allow for drainage to prevent mold growth. Water the trays from the bottom to restrict the amount of water that comes into touch with the greens. It's also a good idea to open the window now and again to improve air circulation and avoid excess wetness.

Step 5. Microgreen's harvest: After germination, microgreens may be ready to harvest in a few days or a few weeks, depending on the species and growing conditions. Microgreens can be harvested with a clean, sharp knife or scissors shortly before being used in a recipe. Microgreens should always be washed in fresh drinking water before consumption; nevertheless, because the plant tissues of microgreens are highly delicate, their shelf life may be significantly reduced after washing. Microgreens might be bagged and stored at low temperatures, then washed right before use as an alternative. If your growing medium is organic, you can compost it after harvesting your microgreens, while your planting trays or pots can be washed and reused.

Storage and Commercial Transport

If stored and packaged appropriately, microgreens have a long shelf life. The best ways to store and transport microgreens are currently being researched. Most commercial microgreens are maintained in plastic clamshell containers, which do not provide the proper mix of oxygen and carbon dioxide for live greens to breathe. Microgreens are typically sold sliced and placed in a non-breathable container because of this. Microgreens have a long shelf life when they are newly cut. To improve shelf life, some growers place a food-safe moisture wicking substance in the bottom of the container. When it comes to the longevity of packaged microgreens and their transportation, moisture management and temperature are the two most critical elements. " Microgreens should be kept as cool as possible after harvesting to avoid harming the product. Keeping the goods at room temperature for as little period as possible should be a requirement. To avoid microbial growth, microgreens should be stored in a clean, refrigerated environment at 4°C or less. In a temperature-controlled setting, a thermometer should be mounted and checked on a regular basis. Microgreens should also be kept in a cool, dry place to avoid spoilage. Stacking too many clamshell packets or bags of goods on top of each other might damage the product at the bottom of the stacks, increasing the danger of the final product deteriorating faster. While shelf life varies by species and variety, with proper handling and storage, most microgreens have a shelf life of 5 to 10 days. The product's high level of care, attention, and correct storage must be transmitted to the distribution stages. To avoid contamination and microbial growth, microgreens must be carried in refrigerated rooms. Although it is the responsibility of the end user to observe basic food safety guidelines, growers should instruct customers on how to handle and store the product when the opportunity arises. Growers can achieve the best results for their business goals by employing the best practices and effective crop management outlined in this factsheet. Buckwheat microgreens packaged in films with an oxygen transmission rate of 225 cubic centimeters per square inch per day had a fresher appearance and higher cell membrane integrity than those packaged in the other films examined, according to the researchers. According to the authors, by following these measures, the team was able to retain acceptable buckwheat microgreen quality for more than fourteen days, which is a major extension. LEDs, or light-emitting diodes, now allow researchers to study the effects of narrow-band light wavelengths on seedling physiology. In plant physiology, the carotenoid zeaxanthin is thought to be a blue light sensor. The impact of short-duration blue light on phytochemical components that contribute to the nutritional content of sprouting broccoli microgreens was investigated in a study. Broccoli microgreens were grown on growth pads in a controlled atmosphere utilizing LEDs. Short-duration blue light influenced the nutritional value of broccoli microgreens by increasing key phytochemical components.

Best Varieties of Microgreens

Microgreens are available in a wide variety of varieties. Some types are more difficult for novices than others, therefore here's a list of the top five best to start with:

Radishes: Radishes are a great place to start with microgreens. They germinate quickly (1 to 2 days) and grow quickly in both warm and chilly environments. They're usually ready to harvest in 5 to 10 days. Radish microgreens are strong in vitamin C, but they also have a wide range of vitamins and nutrients. They also contain a significant amount of protein. Radish microgreens do not require pre-soaking and prefer to be cultivated hydroponically; however, they will grow in soil as well.

Figure 3: Radishes.

Source: https://grocycle.com/how-to-grow-microgreens/

Broccoli: Broccoli microgreens are one of the healthiest types of broccolis available. Broccoli microgreens, like radishes, are relatively simple to grow. Even if you merely rinse them with water a couple of times per day, they'll thrive in a well-ventilated jar. However, if you want to get exceptionally big yields, spread them thickly in the soil.

Figure 4: Broccoli.

Source: https://grocycle.com/how-to-grow-microgreens/

Pea: Pea shoots are crisp and sweet, which appeals to children. They have the same flavor as sensitive young snow peas. After soaking for up to 24 hours, pea seeds should be sown at a high density. When they reach around 5 centimeters in length, they should be collected. Vitamin C, vitamin B1, niacin, iron, magnesium, and zinc are all abundant in pea shoots. They're delicious either cooked or raw. If you're going to grow them outside, you'll need to protect them in some way. They're so delectable that birds and other creatures will most likely come to eat them as well.

Figure 5: Peas.

Source: https://grocycle.com/how-to-grow-microgreens/

Mustard: For thousands of years, people have grown mustard. It's usually cultivated for the seeds, but you can also plant it for the leaves, which you can use in salads or stir fries. Mustard microgreens have a pronounced mustard flavor and are vivid lime green in color. You can pick them as microgreens or let them to grow a little longer for salad baby greens. They give salad combinations a great peppery zing.

Figure 6: Mustard.

Source: https://grocycle.com/how-to-grow-microgreens/

Rocket: Arugula is another name for rocket. It's been a popular salad since ancient times, and it's used to give softer greens a bit more of a bitter spicy flavor. Rocket like cool weather and can sprout at temperatures as low as 5 degrees Celsius, making it ideal for growing even in the autumn or winter. Rocket seeds should not be sown too tightly or they may rot. Because the stems of these microgreens are delicate, you must be careful when watering them. Because they bruise readily, harvesting and washing must be done with caution.

Figure 7: Rocket.

Source: https://grocycle.com/how-to-grow-microgreens/

Soaking Microgreen Seeds

It's not difficult to soak your microgreen seeds beforehand. Simply count out the quantity of seeds you'll need. To remove any residue from your seeds, rinse them and drain them in a strainer or colander. Then, pour enough water into a basin to cover the seeds with an inch or more of water above them. Some seeds, such as wheat grass, absorb a lot of water during this step, so depending on the seeds you're using, you might need even more. Allow your seeds to soak for 12 hours or overnight in your covered bowl. Some seeds germinate better in hotter heat, while others do better in cooler weather. I would soak all seeds in cold water as a rule. You may wish to chill your seeds as they soak, depending on the sort of microgreen seed you're using, especially if it's a hot day. Although seeds do not require light to sprout, I recommend leaving your seeds visible while soaking. You won't forget about them this way. Seeds that have been oversoaked will rot and become unusable. Drain the excess water from your seeds in a strainer after they've soaked. After a quick rinse, drain them one again. After that, you can put your seeds back in the dish where they soaked. You don't want them to dry out, but you also don't want them to be submerged in water. To keep the moisture in, cover your bowl with a damp paper towel or plastic wrap and leave them to sprout for another 12 to 24 hours. You'll want to avoid letting your seeds dry up. It's important to keep them damp, but not wet. This is usually accomplished by spraying them with a spray bottle two or three times per day. After 36 to 48 hours from the time you first water the seeds, you should start to see miniature sprouts coming out of them, depending on the species of seed.

They're ready to go on to their growing medium once they reach this point. Soil or a natural fiber pad are usually used. Ideally, you should use pH-balanced water during the soaking procedure. Although, if you're just getting started, tap water will usually suffice. You can use litmus strips to test your water if you wish to utilize pH balanced water. For most seeds, a pH of slightly less than 6.0 is optimum. If your pH is too high, such as 7.0, add 2 tablespoons of lemon juice per gallon of water (about 3.7 liters) to bring it down to the proper level.

Microgreens Safety Precautions

Microgreens are safe to eat for most people. There are a few things to keep in mind, though.

- Microgreens are cultivated in damp settings and, if not treated properly, can house E. Coli and other bacteria. As a result, it's critical to carefully wash all microgreens before eating them.
- Microgreens can be made from a variety of herbs and vegetables. You may be allergic to a specific vegetable in microgreen form if you have a food allergy to it. If you're sensitive to soybeans, for example, you should avoid soybean microgreens as well.
- Some gluten-containing plants, such as alfalfa, wheatgrass, barley, and buckwheat, are usually safe to eat for gluten-intolerant people since they haven't had a chance to start making gluten yet in their microgreen form.
- In any case, if you have an adverse response after eating a microgreen, stop eating it right away and consult your doctor.
- If you're on a blood thinner like Warfarin, eating a lot of

microgreens could make your medicine ineffective. Because many types of microgreens are high in Vitamin K, which is required for blood clotting, this is the case.

MUSHROOMS

Introduction

Fungi can be found in a variety of shapes, sizes, and colors. Macro-fungi is a broad term for organisms with a visible structure that produces spores, often known as fruiting bodies. Unlike green plants, which have chlorophyll in their leaves to absorb light energy for photosynthesis (the process by which plants convert carbon dioxide and water into organic molecules), mushrooms eat other plant material (the substrate).

Life Cycle of a Mushroom

The key life cycle stages for fungi are as follows:

- **Vegetative growth of the mycelium in the substrate:** When spores released from the gills germinate and mature, they create hyphae, which are the fungi's primary route of vegetative development. These are collectively known as mycelium, as they feed, grow, and eventually form mushrooms (in most species). Mycelium appears as minute threads that resemble the mold that forms on bread from time to time.
- **Reproductive growth when the fruit bodies are formed:** Fruiting bodies or mushrooms vary in appearance depending on the species, but they all have a vertical stalk (stipe) and a head (pileus or cap).
- **Production of spores by the mushroom fruit bodies:** The gills or pores on the underside of the cap are where mushroom spores are formed. Throughout its life, the mushroom generates millions of spores, and the life cycle is repeated each time the spores germinate to form the mycelium.

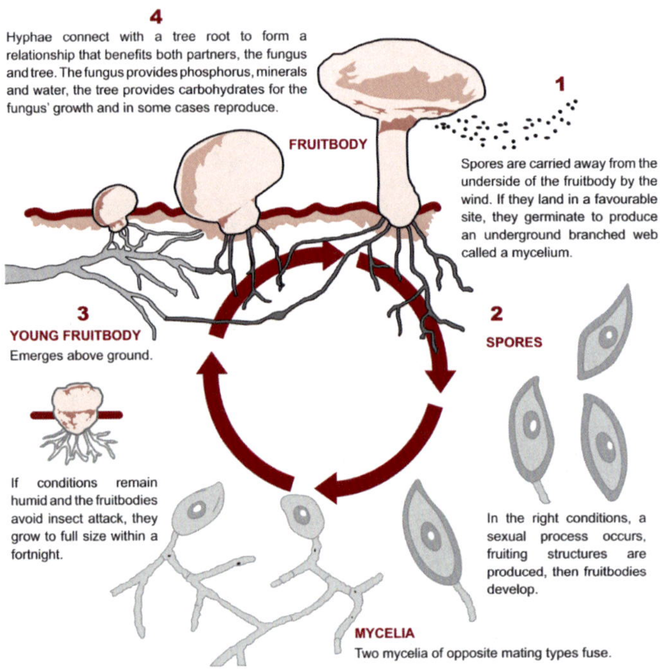

Figure 8: Life cycle of a mushroom.

Source: Marshall & Nair (2009).

Growing Systems

Cultivated mushrooms are fungus that grow on decomposing organic materials and are edible. Mushrooms get their nutrition in one of three ways:

- **Saprobic, growing on dead organic matter:** Saprobic edible fungi can be gathered in the wild, although their cultivated variants are more extensively used as a source of food and medicine. They require a steady supply of sufficient organic materials to maintain production, which can be a limiting issue in the wild.
- **Symbiotic, growing in association with other organisms:** Most wild edible fungi species (e.g., chanterelles - Cantharellus and Amanita species) are symbiotic and commonly form mycorrhizas with trees, in which the fungus aids the tree in collecting water from a larger catchment and delivers nutrients from the soil that the tree cannot access, while the tree provides the fungus with essential carbohydrates.

- **Pathogenic or parasitic:** Plant pathogenic fungus cause plant illnesses, and a small percentage of these micro fungi are consumed as diseased host material.

Essentially, mushroom species can be cultivated in two ways:

- **Composted substrates:** wheat and rice straw, corn cobs, hay, water hyacinth, composted manure, and various other agricultural by-products including coffee husks and banana leaves.
- **Woody substrates:** logs or sawdust.

Although certain species can grow on a variety of substrates, each mushroom species has a preferred growing media. Here, we're concentrating on saprobic organisms. As with truffles, some mushrooms, such as matsutakes and chanterelles, can be grown by inoculating tree roots with species that produce mycorrhizae, which subsequently infect the roots; however, this is not discussed here (see Hall et al, 1998).

Table 4: Key mushroom species and their corresponding cultivation medium

Medium	**Mushroom Species**
Rice straw	Straw (Volvariella), Oyster (Pleurotus), Common (Agaricus)
Wheat straw	Oyster (Pleurotus), Common (Agaricus), Straw (Volvariella), Roundhead (Stropharia)
Coffee pulp	Oyster (Pleurotus), Shiitake (Lentinus)
Sawdust	Shiitake (Lentinus), Oyster (Pleurotus), Lion's Head or Pom Pom (Hericium), Ear (Auricularis), Gano-derma (Reishi), Maitake (Grifola frondosa), Winter (Flammulina)
Sawdust-straw	Oyster (Pleurotus), Roundhead (Stropharia)
Cotton waste from textile industry	Oyster (Pleurotus), Straw (Volvariella)
Cotton seed hulls	Oyster (Pleurotus), Shiitake (Lentinus)
Logs	Nameko (Pholiota), Shiitake (Lentinus), White jelly (Tremella)
Sawdust-rice bran	Nameko (Pholiota), Ear (Auricularis), Shaggy Mane (Coprinus), Winter (Flammulina), Shiitake (Lenti-nus)
Corncobs	Oyster (Pleurotus), Lion's Head or Pom Pom (Heri-cium), Shiitake (Lentinus)
Paper	Oyster (Pleurotus), Roundhead (Stropharia)

Horse manure (fresh or composted)	Common (Agaricus)
Molasses waste from sugar industry	Oyster (Pleurotus)
Water hyacinth/Water lily	Oyster (Pleurotus), Straw (Volvariella)
Oil palm waste	Straw (Volvariella)
Bean straw	Oyster (Pleurotus)
Cotton straw	Oyster (Pleurotus)
Cocoa shell waste	Oyster (Pleurotus)
Coir	Oyster (Pleurotus)
Banana leaves	Straw (Volvariella)
Distillers grain waste	Lion's Head or Pom Pom (Hericium)

Source: Source: Marshall & Nair (2009).

Steps In Mushroom Production

The basic notion in mushroom cultivation is to start with some mushroom spores, which grow into mycelium and expand into a mass large enough in volume and energy to support the formation of fruiting bodies or mushrooms, which is the final step of the mushroom reproductive cycle. The following are the essential generic processes in mushroom production, which can take anywhere from one to three months from start to completion depending on the species:

- **Step 1:** locating and cleaning a specific space or structure where temperature, moisture, and sanitary conditions may be managed in order to grow mushrooms;
- **Step 2:** selecting a growing medium and storing the basic components in a dry, covered area away from the elements;
- **Step 3:** Pasteurizing or sterilizing the medium and bags in which mushrooms will be cultivated, as well as the tables on which they will be grown (to exclude other fungi that would compete for the same space – once the selected fungi has colonized the substrate it can fight off the competition);
- **Step 4:** spawn (spores from mature mushrooms grown on sterile substrate) is used to seed the beds.
- **Step 5:** The most difficult phase is to maintain appropriate temperature, moisture, cleanliness, and other conditions for mycelium growth and fruiting. putting water on the substrate to

increase the moisture level, as this aids in sterilizing;

- **Step 6:** harvesting and eating, or processing, packaging and selling the mushrooms;
- **Step 7:** cleaning the facility and beginning again.

Spawn and Inoculation

Mushroom spawn is obtained from specialized mushroom spawn producers, and each type of mushroom has multiple types or strains of spawn. Because of the care required to maintain spawn quality during the production process, it is not commonly recommended for mushroom growers to generate their own spawn. Spawn is made by inoculating a pasteurized medium, usually grain, with a sterile culture of a certain mushroom species (developed from spores). The cheapest composted substrate cultivation technique is one in which mushrooms are grown in plastic bags holding substrate or compost (which can be sterilized and reused with new substrate) in a basic building to give regulated growing conditions. Bottles can also be used, and wooden trays of various sizes can be stacked in other indoor low-cost methods to give useful cultivating space. Spawn is dispersed completely into the sterilized/pasteurized substrate in an enclosed space under hygienic circumstances to guarantee that the mushroom mycelium grows uniformly throughout the substrate. Farmers with limited finances can avoid having to buy spawn every time they plant a new crop by taking a portion of the substrate colonized by mushroom spawn from the previous crop and utilizing it to spawn the next crop. However, only healthy, uninfected substrate colonized completely by the mushroom spawn should be removed.

Maintaining Suitable Growing Conditions

The inoculated substrate is placed in bags, trays, or other containers and brought to a darkened, enclosed room or building to incubate for up to 12 weeks, depending on the mushroom variety. Plastic bags can be hanging in darkened spaces if space is limited. Water must be accessible and the temperature must be regulated according to the mushroom variety in order for the mycelium to colonize during the next two weeks. The crop should be shielded from direct sunlight and strong winds at all times to avoid drying out. In the growth room, humidity can be maintained by hanging damp rags at various spots around the walls or soaking the floor. A fire (electric if possible) can help manage temperature, and cooling can be aided by blowing a table fan over a pitcher of water. Air passing between the sacks should also help with temperature regulation. To protect the crop from infection, it is critical to maintain sanitary conditions across the entire cropping area.

Harvesting Cultivated Mushrooms

The transition from fully grown mycelium to mushroom fruiting bodies usually necessitates a change in climatic circumstances, such as a decrease in temperature, increased ventilation, and increased humidity. Mushrooms fruit in breaks or flushes, and the type and size of mushrooms harvested are determined by the mushroom variety cultivated and market demand. Mushrooms should be gathered according to market demand; for example, little button mushrooms may command a higher price. Mushrooms are often gathered by hand, with sterilized blades being used to chop the ones that are ready. Pickers should be taught to recognize when the mushrooms are ready to be harvested and to crop them consistently. To limit the danger of harm, handling such a fragile crop should be kept to a minimum.

Marketing Mushrooms

Harvested mushrooms should be handled with care and stored in a container that allows for air circulation, such as a basket, with care taken to avoid bruising. Flies should be kept out of the mushroom baskets, and they should be shielded from the sun, extreme temperatures, and draughts. The best market price is paid for high-quality mushrooms that are healthy and clean. To retain freshness and quality, harvested mushrooms should be delivered to market as soon as possible, or stored in a refrigerated environment, or processed. Both wild and cultivated mushrooms have a tough time getting fresh specimens to market. Fruiting bodies' physical beauty is obviously vital, and client preferences must be respected. If the gills or cap are injured, some species discolor, and they must be handled with caution. To eliminate particles, some preliminary cleaning of gills and gaps may be required, depending on the soil where the fungus grow. It's critical to select fruiting bodies at the right stage of development. Some species become woody and less appealing as they mature, while others decay away.

Pest and Disease Management

The most important aspect of protecting the mushroom crop from pests and illnesses is prevention, which may be achieved mostly through adequate cleanliness. Because mushrooms are typically cultivated in an enclosed setting, the risk of pests and diseases spreading quickly within the crop is significant. It is critical to monitor the crop for pests and diseases on a daily basis to avoid losing at least some of the harvest. Sterilization of the growth chamber and preparation rooms should be done on a regular basis. Control measures should be implemented as soon as pests or diseases are discovered. This may entail carefully pulling affected mushrooms off without spreading the illness, followed by the application of a pesticide. The sort of pesticide necessary should be carefully selected from a list of registered compounds and applied according to the label's instructions. The National Sustainable Agriculture Information Service (ATTRA) and the FAO have more information on Integrated Pest Management.

Scale of production

Growing systems that are most suited to local conditions and depending on the assets available should be chosen. Many mushroom species can be effectively cultivated on a small scale by farmers and other growers with limited resources and who are prone to risk. When growers have developed sufficient skills and income, they can progressively change from a low-cost method to a higher-cost production process with increased output.

Species Selection

Although there has been a lot of research on mushrooms and their growth in temperate areas, there has been surprisingly little research into tropical climate-friendly types. Many commercial mushrooms only fruit at temperatures around 20 °C, making them unsuitable for tropical climates. Tropical strains are more difficult to come by, although some commercial strains that ripen at higher temps can be ordered, and local spore laboratories will be best situated to advise on acceptable kinds and optimal planting procedures. When choosing a species to cultivate, keep the following factors in mind:

- **Availability of waste materials to use as a growth medium:** Not all mushrooms grow well in the same medium.
- **Environmental conditions:** Temperature and other environmental variables are varied for different animals.
- **Available expertise:** Some mushrooms are more difficult to cultivate than others, therefore farmers should start with easy species like oyster (Pleurotus species), which grow on a variety of soils and are suitable for beginners; other options are shiitake (Lentinus edodes) and maitake (Grifola frondosa).

- **Available resources:** It is vital to determine what equipment is required and/or already exists.
- **Market demand:** If producers wish to trade.

Profitable Mushroom Species

A few species of regularly grown edible mushrooms with global significance are listed below.

- **Agaricus bisporus (The white button):** The white button mushroom is the most widely farmed fungus on the planet, and it is especially important in temperate climates. It is often cultivated in higher technology systems and grows in composted substrate. It requires a low temperature of 14 to 18oC to offer optimal fruiting conditions for the mushroom and for the greatest results in cultivation. The button mushroom is the most cultivated fungus, yet it is a temperate species that does not thrive in warmer climes. Furthermore, because all fungus prefers moist surroundings, humidity must be maintained in drier climes. Button mushrooms could be grown along the coasts of hot countries, and some have been grown in abandoned mine shafts in Namibia (to avoid water tainted with heavy metals or other pollutants), where temperatures are low and humidity is high. Tropical mushrooms, such as the oyster mushroom, which is the world's second most cultivated mushroom, thrive in hot climes. Oyster and shiitake mushrooms thrive in high temperatures and thrive in Namibia if humidity is maintained.

Figure 9: White Button (Agaricus bisporus).

Source: https://www.mushroomcouncil.com/varieties/white-button/

- **Pleurotus ostreatus (Oyster):** Because oyster mushrooms are easier to produce than many other species, they are a suitable choice for beginning farmers. They can also be used as part of a sustainable agriculture system that uses organic waste, can be grown on a small scale with a low initial investment, and convert large amounts of substrate to fruiting bodies, potentially increasing profitability. Oyster mushrooms were originally grown on tree logs, but they are now commonly grown on sawdust, wheat or rice straw, and a variety of other high-cellulose waste materials, reducing the fruiting period to about two months. The sterilized and inoculated substrate is simply placed in plastic bags and kept cool and dark during cultivation. After the mycelium has spread throughout the substrate, the bag is cut open to allow fruiting bodies to form. They do, however, have some disadvantages. These mushrooms have a soft and delicate structure; they have the shortest shelf-life of any cultivated mushroom, often showing signs of bacterial or fungal contamination within a day or two of arriving at the market. Some people are allergic to the spores that are released in large quantities when the fruiting bodies emerge from the growing bags, necessitating the use of a face mask in production areas (and air cleaning equipment or respirators in more high technological systems).

Figure 10: Oyster Mushrooms (Pleurotus ostreatus).

Source: https://whyfarmit.com/mushrooms-for-profit/

Lentinus edodes (Shiitake): Shiitake mushrooms are an excellent low-input alternative business since they can be farmed on a small scale with a cheap initial expenditure. Shiitake mushrooms are grown on logs outside, or inside and outside on compressed sawdust, in bottles, or in bags. A growth system based on compressed sawdust and bags provides for a considerably shorter fruiting cycle and a higher rate of return, but it necessitates more skilled administration than log production. The narrower the diameter of the substrate logs, the faster fruiting bodies appear, while production lasts less time, and the denser the wood, the longer the production lasts. Logs are inoculated with spawn from an appropriate and locally sourced strain, and must be kept shaded, moist, and out of the wind as the spawn develops and the mycelium grows throughout the log. When the mycelium has completely occupied the logs and the temperature and humidity are conducive to fruiting, the mycelium will form small "pinheads" on the log's surface, which will grow into mushrooms in a matter of days.

Figure 11: Shiitake (Lentinus edodes).

Source: https://whyfarmit.com/mushrooms-for-profit/

Volvariella volvacea (Paddy straw): In parts of Southeast Asia, particularly Vietnam, paddy straw mushroom farming is frequently combined with rice production. Rice straw, cotton waste, dried banana leaves, and oil palm bunch waste are among the substrates where the mushrooms grow, however yields are lower than with paddy straw, where growing procedures are like those of common or oyster mushrooms. Mushroom producers in many rural locations, such as Indonesia and Malaysia, just place wet paddy straw under trees and wait for mushrooms to appear.

Figure 12: Paddy straw (Volvariella volvacea).

Source: https://mushroommountain.com/how-to-grow-paddy-straw-mushrooms-volvariella-volvacea/

Hericium erinaceus (Lion's mane): Bearded tooth fungus, also known as lion's mane mushroom, monkey head mushroom, satyr's beard, bearded hedgehog mushroom, pom pom mushroom, or bearded tooth fungus. The edible mushroom lion's mane belongs to the tooth fungus family. Long spines (more than 1 cm long), prevalence on hardwoods, and inclination to produce a single clump of dangling spines distinguish it as a native of North America, Europe, and Asia. The fruit bodies can be picked and used in the kitchen. Clinical research has shown no high-quality evidence that the lion's mane mushroom has therapeutic effects. Hericium erinaceus is easily confused with other Hericium species that grow in the same area. In the wild, these mushrooms can be found on hardwoods, notably American beech, in the late summer and fall. H. erinaceus is usually classified as saprophytic because it feeds mostly on dead trees. It can, however, be found on living trees, suggesting that it is a tree parasite. This could point to an endophytic environment.

Figure 13: Lion's mane (Hericium erinaceus).

Source: https://whyfarmit.com/mushrooms-for-profit/

Morchella esculenta (Morel Mushrooms): Morchella, or true morels, is a genus of edible sac mushrooms in the order Pezizales that is closely linked to anatomically simpler cup fungi (division Ascomycota). The network of ridges and pits that make up the caps of these unusual fungi give them a honeycomb appearance. Gourmet cooks admire morels, especially in French cuisine. Commercial harvesting of wild morels has become a multimillion-dollar industry in the temperate Northern Hemisphere, particularly in North America, Turkey, China, the Himalayas, India, Pakistan, and Kashmir, where these highly sought fungi are found in plenty, due to problems in cultivation. Morchella esculenta was first described in 1794, and the genus has been the subject of much taxonomic debate over the years, particularly over the number of species involved, with some mycologists recognizing as few as three and others over thirty. Morchella is thought to have around seventy species globally, with most of them demonstrating strong continental endemism and provincialism, according to current molecular phylogenetics. Several new species have been described from Australia, Canada, Cyprus, Israel, Spain, and Turkey, and the genus is currently the subject of extensive phylogenetic, biogeographical, taxonomic, and nomenclatural research.

Figure 14: Morel Mushrooms (Morchella esculenta).

Source: https://whyfarmit.com/mushrooms-for-profit/

Pros & Cons of Growing Mushrooms for Profits

Mushroom cultivation may be a lucrative and satisfying enterprise. Before getting started, aspiring growers must understand both sides of the equation: the benefits and challenges of cultivating mushrooms for profit.

Mushroom farming pros:

- Low overhead costs
- Little experience required
- Mushrooms can be sold fresh or preserved for future sales
- The specialty mushroom industry is growing
- Less labor-intensive than other types of farming

Mushroom farming cons:

- Time-consuming
- Controlling bacteria and disease can be difficult
- Fresh mushrooms spoil quickly
- Specialized equipment is required at larger scales (sterilizer, fruiting chamber)

SOFT FRUIT / FRUIT TREES SMALL GARDENS

Introduction

There are many suggestions for producing tasty berries and soft fruits in small places; regardless of your gardening situation, you may grow a diverse variety of kinds, and it is also simple. Fruit trees may fit in your borders and grow in containers, so you don't need an orchard, a dedicated allotment, or a greenhouse. Many fruit bushes thrive when they are planted against a sunny wall or fence.

Consideration For A Small-Scale Soft Fruit Growing

Planning a home fruit garden takes a little more foresight than planning a vegetable garden. Most vegetables are annuals or biennials, meaning they only stay in the garden for one or two seasons. Most fruits are trees or shrubs that can live for up to 50 years. Making a list of what you want to eat is the first step in creating an edible garden, whether it's a fruit or vegetable garden. Grow foods that you and your family enjoy. Make a list of everything you want to do. After that, figure out how much of each crop you'll consume fresh, cooked, or preserved. Estimating each plant's production can assist you figure out how much of each crop to plant.

Size of the Fruit Garden:

The size of your garden will determine how much fruit you can grow.

- Strawberries, blueberries, and raspberries, for example, require only a planting bed that is 3 feet wide by 6 to 10 feet long, if that. A strawberry plant will live for three years in the garden; a raspberry or blackberry plant will live for five to ten years or more; and a blueberry plant will live for 40 or 50 years. A dwarf or trained fruit tree, such as an apple or pear, will require a garden space of around 4 feet by 4 feet. A semi-dwarf apple tree will take up almost twice as much space. Fruit trees grown in the wild grow to be much larger—25 feet by 25 feet or more—and may be too enormous for a home garden.
- A dwarf or semi-dwarf fruit tree can survive 15 to 20 years, while a regular fruit tree can live anywhere from 40 to 100 years, depending on the variety.
- A fruit tree trained as a column may need as little as 3 feet by 3

feet of area. Fruit trees with spurs, such as apples and pears, are ideal for columnar growth.

- Many fruit trees can be planted as cordons or espaliers, requiring only a 3 foot by 3 foot or 3 foot by 6 foot planting bed against a post, wall, fence, or house. Cordons and espaliers work nicely with apples and pears. Plums and cherries can be trained to grow on a trellis or espalier as well.
- When it comes to planting fruits, time is important as well as space; most soft fruits will give fruit the same year or in the spring after they are planted. A juvenile fruit tree will produce fruit in three to four years, whereas a mature fruit tree may take five to six years to produce fruit. (Of fact, many fruit trees will continue to give fruit for 40 or 50 years or more after they start bearing fruit.)

Site and Soil for Growing Fruit:

- Plant soft fruits and fruit trees where they will receive full sun for the entire day, ideally 8 hours or more.
- Fruits should be planted in loamy, compost-rich soil that is deep and well-drained. Planting fruits where there is standing water after a downpour is not a good idea. When the soil is loamy and the trees are well nourished, the roots of fruit trees develop deep. Strong trees are supported by deep roots.
- Plant fruits where they will be protected from the wind or breeze. A breeze can dry out leaves and fruit, while wind can do the same and even damage branches.
- Planting fruits in low places where cold air and frost can accumulate is a bad idea. Flowers can be damaged or killed by cold weather and frost in the spring, and ripening can be slowed late in the season.
- Before you start planting, make sure you have access to your garden space. Decide where you want your fruits to grow best, and make sure there's enough space for vegetables as well.

Spacing Plants in the Fruit Garden:

- Plant fruits near walls or structures that can capture solar heat during the day and radiate it back out to the garden at night if you live somewhere with chilly summers. In summer gardens that are cold, a south-facing wall will keep fruits warm and fruitful.

- If you reside in an area where spring temperatures fluctuate from warm to chilly, put fruits on the north side of a building or wall where temperatures are constantly cool until spring arrives and temperatures warm evenly. Avoid planting in areas where the temperature fluctuates dramatically from week to week; significant temperature swings might injure or kill fruit blossom buds in the spring (and cause uneven ripening in summer).
- Fruit trees should be planted in areas where other fruits and vegetables won't grow too close. Except for a thick layer of aged compost or leafmold, leave the ground beneath fruit trees naked. Fruits should not be forced to compete for water and nutrients. Fruit trees can be planted in their own garden or alongside the veggie garden. Irrigate deep-rooted fruit plants separately from shallow-rooted vegetables. Weeds will be kept at bay under the fruits if they are mulched. Avoid cultivating weeds mechanically near fruit trees to avoid disturbing or harming the roots.

Space Saving Growing Methods:

- Consider space-saving fruit-growing techniques including cordons, espaliers, stepovers, single-stem or columnar trees, easily moved pots on plant dollies, and hanging baskets.
- Cordon fruit trees have a single stem or trunk with short fruiting lateral branches that grow at 45-degree angles from the trunk; the fruits are picked at harvest height. Cordons are frequently free-standing or wire-supported.
- Espaliered fruit trees have a single trunk as well, but the branches are normally trained at right angles to the ground and grow horizontally. Espaliers are commonly grown against a wall or fence and are supported by a trellis or wires.
- Small fruit trees with low lateral branches trained to a wire at knee height or below are known as step-overs. Stepovers are frequently used as planting bed edging.
- In the garden, columnar fruit trees take up little more than vertical space. Fruit on short spurs grows the length of a single-trunked columnar tree; apples and pears can be taught to a single-stem or

columnar shape. In a 10 by 10-foot space, you can grow four or five distinct columnar fruit trees.

- Fruits planted in containers can be pushed across patios and pathways on heavy-duty plant dollies to catch the sun in limited places. When frost threatens, they can also be moved indoors or under overhangs.

Stepped Planting:

- Fruit trees and rows of fruiting brambles should be planted in areas where they will not cast shadows on other crops. If you want to shade shade-tolerant plants like lettuce and leafy greens, mint, and rhubarb, grow giant fruits to the north of the garden, where shadows will not fall on other delicious crops.
- Using a phased planting approach is a good idea. Plant standards (tall trees that grow to 20 feet or more) in the garden's north end, then semi-dwarf trees (tall trees that grow 12 to 15 feet tall) directly to the south of the taller trees, then dwarf trees and huge bush fruits in the garden's south end (which grow 6 to 10 feet tall). After that, plant espaliers of hard fruits like apples or pears, followed by cordons and trellises of soft fruits like grapes, and finally little bushy soft fruits like currants and raspberries. Finally, place the shortest-growing fruits, such as strawberries, in the garden's extreme southern end, where they will receive plenty of sunlight. This is a stepped planting strategy, with the tallest plants to the north and the smallest plants to the south, with the tallest plants to the north and the smallest plants to the south.

Planting Microclimates:

- Make use of your garden's microclimates. Most gardens include at least two or three microclimates–areas that are warmer or cooler, sunnier or shadier, and have drier or wetter soil than the rest of the garden. Plant a peach tree in the garden's hottest and sunniest area, and an apple tree in a cooler spot. Warm-spots are frequently located on the south side of a wall or near a patio. Dry areas are frequently found on the other side of a wall or fence that provides protection from the wind and rain.

Choosing Fruits for Your Climate:

- Make a list of the fruits that you and your family enjoy. Those should be at the top of your list of fruits to plant in your backyard.
- Check to see if the fruits you want to cultivate will thrive in your climate. Determine your USDA growing zone and the fruits that can be grown in that zone.
- Apples, pears, plums, and cherries, as well as soft fruits like blueberries, raspberries, blackberries, gooseberries, and currants, may be grown in most temperate areas; some varieties can even be grown in warm temperatures.
- Peaches, nectarines, apricots, citrus fruit, melons, and grapes, which are native to warm Mediterranean climes, will thrive when winters are largely moderate.
- Avocado, guava, and papaya trees, for example, are native to subtropical and tropical climates and can only be grown outside where winters are mild and freezing weather is not a hazard.
- If the fruit you wish to produce doesn't grow naturally where you live, your only option is to grow it in a greenhouse or plastic hoop house heated by solar, gas, or electricity. Consider the additional cost of heating that area.

Fruits That Need Cold:

- Check to verify if the fruit you intend to produce requires winter dormancy. Some plants, such as many fruit trees and bushes, benefit from winter dormancy because it allows them to relax and preserve energy for the following growing season. Some fruits necessitate a period of chilling. Chill hours are the total number of hours in which the temperature is between 34° and 48°F; these hours are required for some fruit trees to blossom and set fruit in the spring. Some apples require 900 chill hours to flower and set fruit, so if you reside at the Gulf Coast or in southern California, you might not be able to grow the kind you want. Check with a local garden center or the Cooperative Extension Service in your area to see if the fruit type you wish to cultivate will grow in your garden and if the winters are cold enough for trees to go dormant.
- Don't give up if the temperature or winters are too warm or too cold to grow the fruit you want to plant. Most fruits have cultivars

and variants that can grow in cooler or warmer climates than other fruit varieties. Many cultivars have been produced to thrive in areas where they aren't native, provided the growing method, circumstances, and microclimates are ideal. In New England, peaches, melons, and grapes will be grown, while in Florida, apples will be grown. Before you decide that a fruit won't grow where you live, investigate the growing requirements of specific varieties. Most fruits come in dozens of different types.

Containers for Tropical Fruits:

- Rather of growing in the open yard, tender subtropical and tropical fruits can be cultivated in pots. Container-grown figs and citrus fruits are excellent possibilities. Fruits that are sensitive to frost can be cultivated in containers and transferred indoors or undercover in the winter; in the summer, they can be returned to the garden. If you're going to grow fruit trees in containers, look for dwarfing kinds (use an all-purpose loam-based potting mix with horticultural sand added for good drainage).

Pollination and Self-Fruitful Fruits:

- Determine whether the fruit variety you intend to grow can produce fruit on its own or requires a pollinator, such as a second tree or plant of the same or a different type, to pollinate and produce fruit. Choose a self-fruitful variety that does not require a mate if space is restricted.
- If space is limited, choose a single-trunked tree with numerous fruit kinds grafted to a single rootstock. Several different apple or cherry varieties may coexist on a single grafted plant. Grafted plants frequently include varieties that aid in pollination.
- Look over the fence or around the neighborhood to see if any fruit trees that are like the one you wish to grow are growing nearby; honeybees and other pollinators can go from one plant to the next to promote pollination.

Planning for Harvest:

- Pick fruit-bearing plants that are simple to harvest. If you don't have a tall ladder or are terrified of heights, don't grow a 25-foot-tall regular size apple. Choose a semi-dwarf or dwarf tree that is easy to harvest by hand or with a fruit picker while keeping your feet planted firmly on the ground.

- Find out when the fruits you wish to grow will be ready to pick. Early in the season, halfway, and late in the season, some fruits are ripe for harvest. Grow an early season apple tree, a mid-season variety, and a late-season variety for a long harvest. For a continuous supply from spring to fall, plant June bearing, everbearing, and day-neutral strawberries.
- Learn how to use the fruit from the plant you wish to produce, whether it's for fresh consumption, cooking, or preserving. Some apples should be eaten raw, while others should be cooked or preserved. Choose kinds that are appropriate for the purpose you have in mind. Pears and cherries are in the same boat.

Planting Soft fruit/Fruit Trees

Soft fruit may thrive in a variety of soil types, although it prefers to be rich and well-drained. They don't like heavy, saturated soils, so if you have clay soil, dig in a lot of organic debris. Before planting, loosen the soil by digging it over and adding a handful of fertilizer. Fruit bushes cultivated in containers can be planted at any time. However, I recommend attempting bare-root plants, which must be planted in the fall or winter while dormant and are far less expensive to purchase. Before transplanting, make sure the roots have been soaked. New plants should never be planted too deeply. This can destroy the plant by suffocating the roots. As a guide, use the 'soil mark' on the stems. Some of these plants will need to be pruned as soon as they are planted. This is critical for them to survive the journey and create healthy new growth, so don't skip it. During dry seasons, keep new plants well-watered for the first several months until their root systems form.

Growing Fruit & Berries on the Patio

Most soft fruit shrubs and plants, if not all, may be cultivated in pots. This has the advantage of making it easy to safeguard ripening fruits from birds. In a short space, a variety of sorts and variations can be planted to create a decoratively distinct element for a balcony, patio, or sidewalk. Soft fruit bushes like Gooseberry, Blackcurrant, Redcurrant, and White-currant can be grown in separate 12-liter containers for a long time. Blackberries require a larger container - 15 liters is acceptable, while Loganberries require a container of comparable size. Strawberry plants can be cultivated in a variety of pots, hanging baskets, grow bags, and troughs. If necessary, blackcurrant, blackberry, and Gooseberry can be cultivated in a shadier location, but strawberries, red, and white currants require plenty of sunlight to yield flavorful berries. Blueberries grow well in containers and are very appealing, but make sure you use a compost that is free of lime. They're frequently advised for use in shrubbery since they look as handsome and beautiful as any ornamental shrub while also providing a bounty of tasty berries. When planted in containers, you won't need to prune these fruits as often because they don't create as much new growth, making them excellent for hands-off gardeners. Cut out any dead, old, or unproductive wood each Autumn, and then give them a healthy feed with liquid fertilizer or a light dusting of potash the following Spring to help them get through the major growing season. Watering on a regular basis is essential, especially as the fruits grow; otherwise, the berries may become dry or too pippy. Fruit trees can be kept in these pots for 3 to 4 years; after that, re-pot in the same size container after teasing out matted roots and shaking out some of the old compost, so you can re-pot with fresh soil. Never use top soil and always use a loam-based potting compost. Strawberries are the lone exception, as they require peat-based compost. If you don't have space for a dedicated fruit growing area and are growing fruit among the flowers and in mixed borders, as is so common these days, this is a terrific option. The habit is similar to that of a little standard rose, with the fruit bush sitting on top of a clean straight stem grafted at a height of around 100cm. Not only do these look better than a regular ground level bush, but they're also easier to harvest and manage, and you can grow smaller plants beneath them, such as herbs, annuals, or bulbs, so they don't take up much room. Plant them at least 48" apart; you can also keep them in 10-liter containers to line a route or road if necessary. As standards, you can cultivate gooseberry, blackcurrant, red, and white currants.

Growing as Cordons

It's simple and popular. You can plant a wide variety of gooseberries, redcurrants, and white-currants as space-saving cordons if you have a spare fence or wall, or simply the space for a couple of posts and wires. You can grow single cordons in a 'V' shape with two main stems, or you can grow double cordons in a 'V' shape with two main stems. When compared to a bush that can be 5-6' across at maturity, you only need to plant 18" apart for a single cordon and 24" apart for doubles, which saves a lot of room. Because the sun is scattered more equally over the plant, the berries ripen early and evenly. Gooseberries thrive on south or west facing walls and fences, although they can also thrive on east or even north facing walls and fences.

Berry Plants as Ground Cover

If you have an unused area, even if it is in partial shadow, you can use it to grow berries as ground cover as long as the soil is reasonable and can be well cultivated. Strawberries thrive surprisingly well in this informal cultivation strategy, thanks to their traveling runners. You can also utilize blackberries, which will simply romp about without assistance. Cranberries, like Tayberries, are good because of their modest spreading growth. None of these fruits require direct sunlight to thrive.

Soft Fruit that are easy to Grow

To cultivate your own fruit at home, you don't need an orchard. In a British garden, apple trees, strawberries, rhubarb, and figs will all thrive. If you don't have much space, try growing your fruit in containers. Strawberries can also be grown in hanging baskets.

Strawberries: The fresh, juicy flavor of sun-warmed strawberries plucked straight from the garden is loved by all. They're so adaptable that they can be grown in pots, hanging baskets, Flower Pouches®, and window boxes, as well as directly in the ground. Grow our 'Extend the Season' pack in June and July to assure a steady supply of delicious, sweet strawberries. Plant strawberries in late summer or early autumn, but find a location where they haven't grown in a few years. For tasty, ripe fruits, choose a warm, sunny location. Dig a hole large enough to accommodate the roots and level the crown with the soil surface. Fill it with soil and water. Plants should be 45cm apart. Every few years, replace strawberry plants and relocate them to a different site.

Figure 15: strawberries.

Source: https://www.daviddomoney.com/how-to-grow-your-own-fruit-and-plant-soft-fruit-bushes/

Raspberries: Autumn fruiting raspberries are self-supporting, so you can grow them in containers or in clumps in your garden. This easy-to-grow crop can be harvested for a delectable dessert from late summer to early fall. They're easy to care for; simply shear the canes to ground level in February and enjoy another luscious crop. There are two sorts of raspberries: floricane (fruiting in the summer) and primocane (fruiting in the winter) (autumn-fruiting). Early in the summer, floricanes produce all their fruit at once. From late summer to the first frosts, primocanes yield a slower crop. Dig a hole large enough for the rootball in a sheltered area in full sun or partial shade. Plant the canes 45 cm apart and backfill with soil. If you don't water your canes well and keep them irrigated during dry seasons, they'll die. Regular canes should be pruned back to 25cm from the ground after planting. Only some types of summer-fruiting canes sold as "long" canes are exempt. The first year after planting, these are ready to bear fruit and should not be clipped.

Figure 16: Raspberries.

Source: https://www.daviddomoney.com/how-to-grow-your-own-fruit-and-plant-soft-fruit-bushes/

Blackberries: In your own garden, forage for hedgerow fruits! Cultivate blackberries in that shady spot behind the shed, or let them scramble over an unattractive fence - it's that easy to grow your own fruit. These tasty fruits may be grown practically anyplace and don't require much care. To make harvesting easier, train the stems onto wires; if you don't want to be pricked, use a thornless type like Apache. Blackberries are ramblers who need to climb up or over something. Attach horizontal wires to a wall or fence panel with trellis or horizontal wires. In full sun or light shade, choose a well-drained, sheltered location.

Figure 17: Blackberries.

Source: https://www.daviddomoney.com/how-to-grow-your-own-fruit-and-plant-soft-fruit-bushes/

Dig a hole large enough for the root ball and plant it slightly deeper than it was previously. Plants should be spaced at least 2 meters apart and well-watered. To encourage vigorous new growth, trim back all stems to 25cm from the ground after planting.

Currants: Redcurrants, blackcurrants, and whitecurrants are soft fruits that work well in sweets, jams, and jellies, and sauces. They also freeze nicely, so you may enjoy the taste of summer even in the dead of winter. If you're short on room, try our Redcurrant 'Rovada,' which is trained as a cordon and is hence ideal for container gardening. Blackcurrants, redcurrants, and whitecurrants are all excellent garden fruits. They don't appreciate strong winds or late-spring frosts, which can damage new leaves, so choose a very sheltered location. Water well after planting in moisture-retentive soil. Plants should be spaced 1.2 meters apart. Cut the blackcurrant branches back to 10cm above ground level after planting. Cut the red and white currants' branches in half.

Figure 18: Currants.

Source: https://www.daviddomoney.com/how-to-grow-your-own-fruit-and-plant-soft-fruit-bushes/

Blueberries: If you want to produce fruit in containers, blueberries are a good choice. What more could you ask from a plant than fragrant blossoms in the spring, flame colored autumn foliage, and nourishing crops of fresh blueberries in the late summer? All they need is acid (ericaceous) soil, which may be purchased at your local garden center. Blueberries are a low-maintenance plant that bears fruit after around three years and makes a lovely patio plant in the meanwhile. If you're short on space, try the Blueberry 'Top Hat' variety. Rainwater is preferable to tap water for watering blueberries because the lime in tap water reduces soil acidity over time. Blueberries are unique among soft fruits in that they require acidic soil to thrive. They'll struggle in typical garden soil, so grow them in ericaceous compost-filled containers. Plant to the same depth in a container with plenty of area for the root-ball. Rainwater is preferable than tap water, which is too alkaline for acid-loving plants.

Figure 19: Blueberries.

Source: https://www.daviddomoney.com/how-to-grow-your-own-fruit-and-plant-soft-fruit-bushes/

Gooseberries: Gooseberries are adaptable enough to be harvested early for savory sauces, or left on the bush to ripen and sweeten for cakes, crumbles, and cordials - or eaten fresh and deliciously right from the bush. For red-skinned luscious sweetness, choose 'Xenia,' or 'Invicta,' which produces large cropping bushes with fruit that is great for cooking. Gooseberries require deep soil that is not very dry in the summer. In full sun or partial shade, use a moist-retentive location. Plant the canes somewhat deeper than they were before and give them plenty of water. Half-prune the canes after they've been planted.

Figure 20: Gooseberries

Source: https://www.daviddomoney.com/how-to-grow-your-own-fruit-and-plant-soft-fruit-bushes/

MICRO DAIRIES

Introduction

A micro-dairy is described as a dairy farm with less than 40 cows, with some referring to those with 1-5 cows as nano-dairies. Although the word is not formally defined, these values represent current practitioners' broad agreement. Other elements that may be considered when defining a micro-dairy include labor input for cultivating the dairy, pasture land, and the number of liters of milk sold. Again, there are no specific definitions for any of these indicators, nor are there any numerical constraints, although the numbers tend to be proportionate to the size of the dairy. We'll be looking at micro-dairies through the lens of herd size in this paper.

Micro-dairy Business Model

The benefits of the micro-dairy business model are derived from the reduction in herd size. This decrease in size is coupled by a decrease in production costs due to the use of fewer materials and more grass, usually with more work. Importantly, the farmer's price is boosted, usually by selling retail rather than wholesale. Farmers keep a far bigger percentage of the milk's value when they sell directly to consumers or into local shops. When sold retail, the milk produced by an average dairy cow is worth between £6,000 and £14,000, compared to £2000 if sold to a processor. Following the micro-dairy strategy, a herd of 20 cows may generate over £100,000 in revenue. Other aspects of the supply chain are included in the business model, which works best with a network of small herds feeding into a small processing plant, followed by delivery rounds and sales to stores. Producers can form a cooperative to invest in processing and other aspects of the supply chain, although processing does not have to be done by the same person as the producer. The strategy works because the consumer does not pay much more for the goods, if at all, but a large portion of what they do pay goes to the farmer. Some consumers or processors willing to pay a premium for a high-quality product – fresh, tasty, Agri-ecologically produced milk given by the farmer or someone close by – might help generate additional cash. The price paid by consumers can be kept in the business by cutting out the intermediaries and maintaining the supply chain as close to the farm as possible. With 40 cows, you can make a fair living rather than fighting to make ends meet with 400.

Strategies for Profitability for Micro Dairy Producers

With long-term average all-milk prices ranging from $12.00 to $13.25 per hundredweight, each dairy producer must determine whether to stay in the business. If someone is now experiencing too low profitability and is hoping that farm-level pricing will improve on their own or with increased federal dairy policy support, they must accept the fact that higher milk prices are unlikely. If they are unable to lower production costs, they may have little choice but to abandon dairying. Without a doubt, the whole expenses of production for the larger modern dairy facilities that are being built and efficiently managed are far under $12.00 per hundredweight, with some even around $10.00 per hundredweight. These aren't all 1,000-to-3,000-milk-cow operations. It suggests that businesses with 300 to 600 cows can be cost competitive. This company's size allows them to fully exploit technology while also achieving labor efficiencies. Additional environmental regulations will undoubtedly be implemented. These new laws will include animal waste management. Environmental laws will receive a lot of attention during the 1995 Farm Bill debates. Compliance costs are likely to be lower for larger herds than for smaller herds on a per cow or per hundredweight of milk basis. Simply put, building a waste management system for 1000 cows does not cost 20 times as much as building one for 50 cows. All this debate boils down to the reality that smaller dairy herds must think about ways to lower milk production expenses per hundredweight of milk produced. Even so, a certain minimum herd size will be required to generate sufficient net income to support a family. With herds of less than 40 or 50 milk cows, this will be incredibly tough. Even if their per hundredweight production costs are close to the most efficient producers, herds smaller than this will not have enough livestock units to earn appropriate income. To complement dairy income, either some off-farm income or additional income from other farm activities will be required. There may be specialized markets for these smaller herds to investigate, such as going organic and processing the milk into higher-value organic dairy products. A collection of smaller dairy producers in Wisconsin, usually with 35 to 40 milk cows, founded a cooperative that markets organic cheeses, butter, yogurt, and beverage milk not just in the Upper Midwest, but also in the Northeast and even the West. These organic dairy products sell for more than comparable nonorganic products, with organic dairy farmers collecting around $16.00 per hundredweight for their milk. For these smaller herds, and for that matter, all sizes of dairy herds, there are several options for lowering per cow and per hundredweight production costs. The following are examples of these methods:

Maintain a herd in which milk per cow exceeds the state average: According to cost-of-production figures, the most productive herds are not always the most profitable. However, herds with milk per cow that is close to or below the state average are usually not very profitable. Anything less than 17,000 lb of milk per cow in a Holstein herd should be questioned. Smaller herds should consider consulting with nutritionists and herd health specialists. Such expenditures could potentially lower the cost of milk production per hundredweight. More milk per cow could boost the number of pounds of milk produced per work hour while also lowering the fixed cost per hundredweight of milk.

Reduce the capital investment per cow: I don't believe it's possible for producers with smaller herds, perhaps as few as 150 cows, to own a whole range of contemporary equipment from planting to harvesting. They just don't have enough land to distribute the investment's fixed costs. Furthermore, managerial time is split thin during planting and harvesting, and often suffers as a result. Smaller herd owners should consider acquiring all or a portion of their forages and feed. Many smaller farmers, in my opinion, should not raise grain and should instead focus on forages. Grain planting and harvesting equipment are no longer required. Other possibilities include hiring custom operators for planting and harvesting, leasing rather than buying equipment, and sharing machinery with other dairy producers in the region (co-ownership or equipment swap).

Contract for the raising of dairy replacements: Large-scale producers are starting to do this, and it could be an excellent alternative for small-scale growers. It may allow for more time to be allocated to the milking herd by freeing up labor and management time.

Reduce input costs through rotational grazing systems: These technologies have the potential to lower milk production costs per hundredweight while also increasing profitability. A rotating grazing system, like anything else, requires adequate management. Turning the cows out to graze isn't the solution. Furthermore, not all dairy farmers have land suitable for grazing systems.

Smaller producers can consider getting together and building a larger and more modern dairy facility: There are several different types of arrangements that can be made. A cooperative structure, a subchapter S corporation, a limited liability company (not permitted in all states), or a partnership arrangement could all be used to own the facility. Each farmer might keep his or her farm and grow and collect forage and grains to feed the combined dairy herds housed and milked in the co-owned dairy complex. Alternatively, the dairy farmers might own the dairy facility, cows, and acreage as a single farming operation. The goal of this dual ownership is to split the costs of building a larger, more modern dairy facility and having a qualified herdsperson oversee the dairy herd. Combining individual herds would result in cheaper fixed costs per cow and per hundredweight of milk than if each producer developed his or her own more sophisticated milking machine.

Practice sound financial management: Financial success requires excellent financial and commercial management, regardless of the size of the company. Profitability must be considered when making decisions. Accurate business and production records are required. Dairy farmers must keep extensive records of their production costs, not just a checkbook and the bare minimum for tax purposes. This is by no means an exhaustive list. However, it does show that there are other ways to make money in micro-dairying.

Structure and System of Starting a Micro-Dairy

On a small dairy farm, milk is the main product from which you make money. If you supply a fluid market (fresh milk, lactose, etc.), milk output and volume will make you more money. If your market is concerned about quality (e.g., those who process milk into cheese, yoghurt, ice cream, butter, and so on), you can profit by combining milk quality and volume. There are two categories of milk quality: compositional and sanitary. The nutritional content of the milk - fat, protein, whey, and so on – is referred to as compositional quality. Pathogens and other dirt in the milk are linked to hygienic quality, which is linked to cow and milk handling on the farm. Whatever market you're aiming for, make sure your product is safe, nutritional, and healthful. Your dairy will also make extra money by selling bull calves, cows that are no longer needed, and heifers that are more than the farm requires. Manure can either be composted and used on the farm, or it can be sold to other farms. It can also be converted into biogas, which can be used to power the farm.

Best breed: This is a difficult question that prospective dairy farmers always ask me. 'It depends,' I say most of the time. Holstein Frisian, Jersey, Guernsey, Ayrshire, and Red Dane are among the dairy breeds offered. Dual-purpose breeds like Brown Swiss and Simmental are also employed in the dairy industry. Each of these breeds has its own set of characteristics and qualities. The type of breed to keep is determined by several factors, including the production system, production circumstances and conditions, market demands, breed availability, and farmer inclination.

Reproduction is critical: A cow cannot produce milk unless she has a calf. Every twelve months, without fail, all your cows should calve. There will be no milk to sell if there is no reproduction. You can use a bull or artificial insemination to mate your heifers and cows (AI). The most prevalent method of breeding dairy cows is artificial insemination (AI).

Seasonal production is an option: You can employ a seasonal or all-year calving strategy, depending on your market and feed availability. Most dairies have a year-round production method. During the summer, excess fodder is gathered and stored for use during the lean season (late autumn, winter, and spring). When the dairy is pasture-based, seasonal calving is a suitable alternative. All-year-round production is a good approach for feedlot-based or hybrid systems. In such instance, your mating program should be designed so that 75-80 percent of your cows are in milk at any given moment.

Feeding system: Feed can account for up to 80% of a dairy's total operating expenses. As a result, the component that determines whether you make a profit or a loss is usually feed. You take the animal to the feed in a pasture-based approach. Because the cows graze on their own, the expense of feeding them is cheaper. However, such farms have reduced milk outputs and poor reproductive performance. The fundamental issue with a pasture-based feeding strategy is seasonal pasture growth. There are two possible solutions to this issue:

- a solid forage plan - to even out periods of plenty and periods of deficit; and
- seasonal production – to match feed availability and demand by the enterprise.

Other issues revolve around the pasture's sufficiency and quality. Most of these issues are related to management. Alternatively, you can feed the animal – either in a feedlot or using a zero-graze technique. Animals housed in sedentary systems are given a partial mixed ration (PMR) or a total mixed ration (TMR) in a zero-graze system. Feed availability is not dependent on the season, precision feeding is possible, and milk outputs are higher with this technique. The only issue is the cost of running a farm of this size.

Milking system: If you don't have a good system for harvesting the milk, all of your husbandry techniques and investments will be for naught. You should milk in such a way that as much milk as possible is let down and extracted without causing injury to the factory - your cow and its udder. In addition, your milking equipment should ensure that the milk is of the highest sanitary quality. You can milk your cows by hand or by machine. Hand milking is time consuming (4-5 cows per hour), labor expensive, inefficient, and compromises the sanitary quality of the milk. You may need to switch to mechanical milking as your herd grows, which is faster, more efficient, and cleaner. Fixed (in most situations) or transportable milking machines are available. Bucket machines, trolleys, and bush dairy systems are typical in transportable units. Fixed units can be configured in a variety of ways, referred to as parlor kinds. Tandem, side by side, herringbone, rotational, and robotic systems are among them.

Tips for Running a Micro-Dairy

- Maintain a clean, dry, and comfortable environment for your dairy animals.

- Before you feed and care for yourself, feed, and care for your dairy cows.
- Maintain as little complexity as possible in your facilities and management strategies.
- Protect your animals from any stress that may be avoided.
- Don't overfeed your dairy cows. Treat them with the respect that they deserve as coworkers or teammates.
- You shouldn't be afraid of your dairy cows, and they shouldn't be afraid of you.
- Avoid fads in management or equipment that haven't been demonstrated, especially if they're expensive.
- Dehorn dairy calves as soon as you can feel the horn buds with a cordless gas dehorner.
- Feeding your dairy animals for optimal physical condition and health will result in increased milk output.
- Don't feed your dairy cows just to get the most milk out of them.
- Increase the return on your investment by extending the life of your dairy animals.
- Check your cows and other dairy animals for Leucosis and Johne's disease.
- As quickly as possible, get rid of any dairy animals that have tested positive for Leucosis or Johne's disease.
- Most farmers make a lot more money milking cows than they do milking farmers.
- Without spending any money, learn how to solve difficulties.

SMALL-SCALE CHICKENS & EGGS FARMING

Introduction

Small-scale poultry rearing by households using family labor and, if possible, locally available feed resources is defined as family poultry. The fowl are free to roam the residential compound and locate much of their own food, with the householder providing additional food. Family poultry is rarely the family's primary source of income; rather, it is one among a variety of integrated and complementary farming activities that contribute to the household's total well-being. Poultry provides a significant source of income through the selling of birds and eggs. Consumption on a sporadic basis is a valuable source of protein in the diet. In many countries, poultry also has a crucial socio-cultural role. Poultry farming relies heavily on family labor, and women (who often own and care for the family flock) are the primary benefactors. As extension workers and in immunization programs, women play an essential role in the growth of family poultry industry. The major purpose of family poultry rearing is to generate income. Eggs can give a consistent, albeit small, source of revenue, and live bird sales can provide a more flexible source of cash as needed.

Production Systems

Family poultry are kept under a wide range of conditions, which can be classified into one of four broad production systems:

- free-range extensive;
- backyard extensive;
- semi-intensive; and
- intensive

Free-Range Extensive Systems

In Africa, Asia, and Latin America, 80% of farmers use the first two extensive methods to keep poultry. The birds are not limited in free-range settings and can search for food across a large region. There may be rudimentary shelters available, which may or may not be utilized. The birds may roost outside in trees and lay their eggs in the bush. The flock is made up of birds of various species and ages.

Backyard Extensive Systems

Poultry are kept in cages at night but are free to roam during the day. To supplement scavenging, they are normally fed a handful of grain in the morning and evening.

Semi-Intensive Systems

Birds are confined to a specific area with access to shelter in this hybrid of extensive and intensive systems. They can be found in both metropolitan and peri-urban areas, as well as rural areas. The birds are confined in an enclosed area outside during the day and housed at night in the "run" system. To avoid wasting due to rain, wind, and wild animals, feed and water are provided within the house. There are two different forms of housing under the European method of free-range chicken husbandry. The first is the "ark" system, in which the chickens are confined overnight (for predator protection) in a building attached on two rails or skids (typically wooden), allowing it to be moved from place to place using draught power. A average size is 2 2.5 m, which can contain approximately 40 birds. The "fold" unit is the second form of housing, with a stock density (space allowance) for adult birds of 3 to 4 birds per square meter (birds/m2), both inside and (at least this) outside. Typically, the fold unit is small enough to be moved by one person. In poor countries, neither of these two methods is widely used.

Intensive Systems

These systems are utilized by medium to large-scale commercial enterprises, as well as by individuals. Birds are kept in complete confinement, either in houses or cages. The initial investment is more, and the birds are completely reliant on their owners for all their needs; nevertheless, productivity is higher. Intensive systems are divided into three categories:

- **Deep litter system:** Within a house, birds are entirely contained (with a floor space allowance of 3 to 4 birds/m2) but are free to move around. Grain husks (maize or rice), straw, wood shavings, or another absorbent (but non-toxic) substance is used to cover the floor (a 5 to 10 cm deep layer). The totally enclosed system protects the birds from thieves and predators and is appropriate for only a few commercial egg or meat-producing poultry breeds (layers, breeder flocks and broilers).
- **Slatted floor system:** Instead of thick litter, wire or wooden slatted floors are utilized, allowing stocking rates to be increased to five birds per m2 of floor space. Birds are permitted to wander around freely and have less contact with feces.
- **Battery cage system:** This is commonly employed for laying birds who are maintained in cages throughout their productive lives. The method is largely limited to large-scale commercial egg layer operations due to the hefty initial capital expenditure. Commercially intensive systems of raising indigenous chickens are uncommon.

Steps to Establish Small-Scale Poultry Farm

Step 1. Assess your goals and resources: Begin your poultry operation by evaluating your personal business goals (income, time availability, lifestyle, etc.) and making a list of your available business resources (your labor and that of your family or employees, cash, skills, existing structures and systems, waste products that can be used profitably, etc.). Establishing a solid foundation upon which to construct your organization and guide all of your business decisions begins with identifying your goals and assets from the beginning.

Step 2. Decide what kinds of poultry products you want to sell, and where you want to sell them: Once you have identified your goals and assets, you must answer three important questions:

- Who are you going to sell your poultry to? The general public, restaurants, wholesalers, retailers, or a combination of outlets are all options. Study your local markets to see what items are available and what new products are needed, assess demand and prices for existing products, and identify viable niche markets to service as you answer this question.
- What markets would you like to sell your poultry goods in? Do you wish to sell directly from your farm, at an off-farm location (such as a farmers' market), or through wholesalers? The scale, scope, and design of your organization will be heavily influenced by the answer to this question.
- Do you wish to sell eggs, meat, or a combination of the two? The answer to this issue (which is mostly determined by the consumers and markets you chose in Questions #1 and #2) will influence the breeds of chickens you grow as well as the design of your housing, pastures, and processing facilities.

Direct marketing is the most effective, profitable, and fulfilling approach to sell your items, according to most producers. You boost your earnings by pocketing the entire retail price of your product rather than allowing the wholesaler (middleman) to take the retail mark-up by establishing direct contact with your customers on your farm and/or at farmers markets. You can also enhance your revenues by adding value to your items by processing them. Along with your eggs, serve ready-made egg salad or seasoned boneless skinless chicken breast beside your whole birds. Consider obtaining organic certification for your farm, as organic price premiums can treble your revenues. Whether you sell retail or wholesale, organic or not, try to sell all your birds before investing time and money in their production. Also, while determining your retail or wholesale price, make sure to include all your production costs, including labor. Many producers fail to compensate themselves for labor, and as a result, they are unable to earn a profit, support their families, or expand their businesses.

Step 3. Develop a budget: Make a list of everything you'll need for your business, including chicks or pullets, brooder equipment, feed, housing, pasture, fencing, feeders and waterers, processing equipment, product transportation, farm stand fees or rent, and personnel. Assign a dollar value to each item; the overall dollar value of the list will determine the amount of cash you'll need to get your business up and running. This budget will assist you in determining the initial size and scope of your operation, based on your objectives and available resources. The budget will also help you figure out how much profit you'll need to keep your family and business afloat. Discover inventive ways to barter or trade for products, find them used or for free, or fashion them from fixtures that already exist on your farm as you construct your budget. Each inventive idea will lower your startup expenses while increasing your profits.

Step 4. Choose your breeds: You must choose the proper layer or broiler breed(s) to fit your demands after determining the items you will sell and the starting size of your enterprise. Because they gain weight quickly and have a flavor that buyers are acquainted with, many pastured poultry farmers prefer the Cornish Cross as a meat bird. Purebred breeds generate more meat and eggs than hybrid cross fowl. However, good features that have been "selected out" in favor of meat and egg production, like as foraging abilities, disease resistance, and climate change tolerance, are being lost in these younger varieties. Every breed has its own distinct traits, and the breed you choose will be determined by your farm location and marketing strategy. Based on usage, the breeds listed below are some suggestions. These recommendations are for purebreds that were raised as dual-purpose birds in the past:

- **Broilers:** Cornish Cross (currently being bred to regain foraging characteristics) is the preferred meat bird, but other suitable purebred meat birds include the Delaware and Java (though they will not reach market weight as quickly).
- **Layers:** Rhode Island Red, Leghorn, and Plymouth Rock lay many eggs as well as the Ancona (beautiful colored eggs) and Minora.

- **Dual Purpose:** Australorp, Barred Plymouth Rock, Delaware, Java, New Hampshire, Rhode Island Red, and White Wyandotte.

Step 5. Design housing and pasture for your birds: Begin by selecting a site or general area where the house and chickens will reside permanently or within which they'll be moved. When choosing a site consider:

- **Direction:** Orient the house toward the south in winter
- **Soil drainage:** Choose high, well drained ground with a south or southeast slope if possible
- **Space requirements:** Provide 2.0 – 2.5 square feet of indoor space per bird if they have outside access for free range.

You can start designing an ideal pasture and house system for your birds once you've chosen a location. You can choose from a variety of "pre-designed" alternate house and grazing systems, or you can create your own. The system you select or design should be compatible with your farm's location and operational objectives. You must incorporate the following aspects in your housing design, regardless of whatever system you choose:

- Ventilation
- Insulation (not always necessary, but usually required in temperate climates)
- Walls
- Roof (needs to be waterproof and include an overhang for passive temperature regulation)
- Nests (only needed for layers: 1 square foot of area with 1 foot of head room)
- Roosts (2" x 2" stock, rounded or beveled, spaced 12 – 15 inches apart: should not be more than 2 feet off floor).

Step 6. Nutrition: Chickens, like all living things, require a well-balanced meal rich in proteins, carbs, minerals, and vitamins. Protein requirements range from 16-20% of total feed weight, depending on the stage of development and productivity of the chickens. Corn, soybeans, and wheat, as well as vitamin and mineral supplements, make up most diets. Because chickens lack teeth, they require grit (tiny stones or oyster shells) to facilitate digestion by the gizzard (an organ placed before the small intestine). Grazing weeds, weed seeds, legumes, grasses, and bugs provide numerous essential nutrients to pastured fowl. The amount of pasture-based nutrients that the birds can eat and assimilate is a point of contention. Chickens, unlike ruminants, lack a multi-compartmented stomach and are unable to digest cellulose properly. As a result, chickens cannot survive only on pasture. Grazing grassland, on the other hand, is extremely beneficial to birds. Free range birds have been discovered to have much more vitamins and omega-3 fatty acids, as well as a lower fat content, according to studies. Many livestock nutritionists are also organic feed suppliers, so they can assist you in developing a suitable ration for your poultry. Feed prices, particularly organic feed, can be quite pricey. As a result, if you have a large flock of birds, you might want to explore growing some or all your own feed. Feed production can help you save money while also making your organization more self-sufficient and long-term.

Step 7. Maintain your flock and keep them healthy: Pastured fowl are often disease and illness resistant. Weather and predators are the most common health and management issues. In most areas, enough protection is necessary to protect your birds from the elements such as cold, rain, strong winds, and heat. In most places, predator protection, such as portable electric poultry fencing, is also necessary. Checking the birds' health, refilling their feed and water supply, and cleaning their housing and pen spaces are all part of daily care. Keep accurate health records for your birds, including their age, immunizations, egg production, and so on, as this information is useful for tracking and resolving potential health issues, determining production costs, and obtaining organic certification.

Step 8. Manage your pastures effectively: Chickens will graze on any type of pasture, and while they prefer legumes to grasses, they will eat the entire field eventually. Start by leveraging pre-existing pastures when planning your pastures, especially if you already have pastured ruminants. Plant a broad mix of forages that develop at different times of the year to improve soil quality and give grazing variation for the birds if you must replant or create new pasture areas. Consider your site characteristics (such as soil type, rainfall, etc.) and production intentions when selecting forages to grow (such as replanting the area in vegetable crop or using it for other livestock to graze). Joel Salatin, author of Pastured Poultry Profits, uses a mix of "grasses and clovers, including native grasses, broadleaves, clovers, chicories, oats, and rye" to create a "permanent polyculture." Salatin also recommends cutting the grass short (a few inches) to allow the birds to consume more food. Another important and helpful aspect of your poultry pasture is fencing. Fencing isn't required, but it protects the birds from most predators (excluding avian predators like hawks) while allowing them to access ample range space. Fencing also aids in the better management and/or rotation of grazing areas for the flock. Portable poultry fencing, commonly known as "poultry netting," is a cost-effective option. To graze a complete flock, most small-scale growers utilize only a few 165-foot rolls of chicken netting and one battery charger.

Step 9. Plan for economic and environmental sustainability: You must review your poultry operation's ecological and economic sustainability on a frequent basis before, during, and after it has been designed and implemented. Recycling resources on your farm decreases the number of inputs that must be obtained from outside sources and restores many of the resources that are lost due to off-farm exports.

General Management for Poultry

Housing and Runs

In the wild, chickens lay eggs on modest nests, perch in trees, and spend most of the day scavenging for food. Chickens spend a significant amount of time scratching to uncover buried food. Poultry in backyard and semi-intensive production systems are frequently contained at night to deter thieves and predators, but poultry in intensive production are completely always confined. To deter theft, some village households keep their few chickens inside the home or even under their bed at night. When given a choice of where to deposit their eggs, hens choose a soft "litter" base and a sufficiently sized (about 30 cm cube) shaded nest with some solitude. Before entering a nest box, chickens normally evaluate several possible locations. They subsequently demonstrate nesting behavior, including a unique protective nest-seeking voice, before sitting and finally laying. When they have laid an egg, they make another form of "pride of achievement" call to proclaim it. In a battery cage house, these calls can also be heard. If perches are supplied, hens will prefer to perch rather than stand on the wire floors most of the day, and most birds will roost on the perches after dark. Perching is a likely survival trait for avoiding night predators. The following are the essential needs for poultry housing:

- space;
- ventilation;
- light; and
- protection (from weather and predators).

Space: Density of Birds per Unit Area

Because the amount of space available dictates the number and type of chickens that can be housed, this is the most significant basic principle in poultry housing. For example, a 6 m by 11 m deep litter house with a stock density of 3 birds/m2 (3.6 ft2/bird) can store 200 laying hens. Stock density was previously assessed in ft2 per bird, which is the inverse of birds per m2 in the metric system, with a conversion factor of 0.0929 m2/ft2 incorporated. Per bird, linear space or perch length is measured in millimeters. Table 1 shows the recommended floor and perching space for the three main species of chicken. A stock density of three to four birds per square meter is ideal for hen groups. More room allows for a wider range of behavior to be expressed. Less room leads to stressed social behavior, which can lead to disease vulnerability and cannibalism, as well as depriving lesser birds of food and perch space. Individual birds require more space than the 22 birds/m2 (0.5 ft2/bird) density typically utilized in commercial laying cages for appropriate behavior and proper exercise. Animal welfare concerns have prompted study on laying cage structures in recent decades, with the goal of creating designs that are better suited to the needs of hens while being cost-effective for production.

Table 5: Requirement of chickens for floor and perch space

Chicken types	Floor Space (birds/m^2)	Floor Space (ft^{2}/bird)	Perch Space (per bird)
Layer	3	3.6	25 cm (10 in)
Dual Purpose	4	2.7	20 cm (8 in)
Meat	4-5	2.1-2.7	15-20 cm (6-8 in)

Source: http://www.fao.org/3/y5169e/y5169e05.htm

Ventilation: air flow

The importance of ventilation in housing cannot be overstated. Cross-ventilation at bird level should be allowed in the form of floor level inlets, open in a direction that allows the prevailing wind to blow across the breadth of the structure if the building has open sides. Even across an open-sided building, an air mass between the side walls of a poultry house resists being transported. The wider a structure is, the more resistant it is to air flow. Because of the natural tendency of air to resist movement, buildings above 8 m (26 ft) broad have a substantially higher challenge. Buildings that rely on natural airflow for ventilation should not be more than 8 meters wide. Heat stress is a major stumbling block to successful manufacturing and can even be fatal. Although birds can resist temperatures below freezing, they cannot withstand temperatures beyond 40 degrees Celsius. This is determined by the current relative humidity. Poultry lack sweat glands and must cool themselves by exhaling water, which is referred to as evaporative cooling. This cooling method does not perform properly when the humidity is too high. Temperatures of 46°C or above are lethal for most chickens, and temperatures of 40°C or higher cause severe stress. To acquire heat in temperate climates, the chicken house can be built to face the rising morning sun. In the tropics, however, an east-west orientation of the building's length helps to reduce direct sunlight exposure. For this reason, building materials like tin and other metals should be avoided, though white paint will reflect up to 70% of incident solar heat radiation. In this case, ventilation concerns in building alignment may win out over solar heat regulation, as cross-flow ventilation necessitates the building's side facing the prevailing wind. Ground cover can also help to prevent heat reflection. If there isn't much air flow or the humidity is high, shade should be given. When there is no shade or when the temperature is high, chickens get heat stressed and irritated, and they may begin to peck at one other. Blood is easily extracted when new pinfeathers are forming (particularly on young stock), which might lead to cannibalism. Heat stress has the following effects:

- a progressive reduction in feed intake as ambient temperature rises;
- an increase in water consumption to lower temperature;
- a progressive reduction in growth rate; and
- disturbances in reproduction (lower egg weight, smaller chicks, reduced sperm concentration and an increased level of abnormal sperm in cocks).

Light: duration and intensity

It is critical to have a well-lit home. Lethargic, sedentary, and unproductive birds are the result of a dark environment. Because birds recognize food by sight, light is essential for feeding. This is particularly crucial for closely managed day-old chicks, who require bright 24-hour lighting for the first week of their lives. In terms of sexual maturity, light is very significant. As the light percentage in the day increases, as it does naturally from mid-winter to mid-summer, sexual maturity in maturing pullets accelerates, allowing them to produce eggs sooner. If the hens are already laying, the increased light proportion will result in more eggs being produced. The converse effect is also true: as the day's light proportion drops (as it does naturally from mid-summer to mid-winter), sexual maturity in growing stock slows, and egg output in laying hens declines. As the difference in the day's daylight fraction decreases less and less as one gets closer to the equator, these impacts become less pronounced.

It is critical to have a well-lit home. Lethargic, sedentary, and unproductive birds are the result of a dark environment. Because birds recognize food by sight, light is essential for feeding. This is particularly crucial for closely managed day-old chicks, who require bright 24-hour lighting for the first week of their lives. In terms of sexual maturity, light is very significant. As the light percentage in the day increases, as it does naturally from mid-winter to mid-summer, sexual maturity in maturing pullets accelerates, allowing them to produce eggs sooner. If the hens are already laying, the increased light proportion will result in more eggs being produced. The converse effect is also true: as the day's light proportion drops (as it does naturally from mid-summer to mid-winter), sexual maturity in growing stock slows, and egg output in laying hens declines. As the difference in the day's daylight fraction decreases less and less as one gets closer to the equator, these impacts become less pronounced.

Protection: shelter sheds and buildings

The type and choice of housing to protect chickens from the effects of weather and predators is influenced by several factors. The local temperature, accessible area, flock size, and management system are all factors to consider. Birds in large systems must be protected from disease and predators while yet being able to forage. Traditional live plant fence for large animals is ineffective against predators including snakes, kites, rats, and other pests. To dissuade predator birds, tie parallel lines of string across the primary scavenging region with intervals fewer than the predator's wingspan; or, alternatively, lay a fishing net held on poles across the side of the run where predator birds could swoop on the scavenging chicks. Large predators can be caught with leg traps. Traps do not need to be set around all the pens because predators often attack the same pen on the second night. Steel traps can be camouflaged and protected by boiling them in walnut hulls or cocoa pods. Because most predators have a great sense of smell, the traps will be more effective if they are not touched with bare hands. Instead, use a stick, rubber gloves, or tongs to handle them. When the birds are young, rats, mongooses, and snakes become a nuisance. Rats frequently enter through the ground flooring, and abnormally silent chicks gathered under the brooder warmer or in a corner, or dead chicks with small bloody neck scratches, may be the first indicators of a rat assault. If snakes get access to the brooder house, they will kill the chicks. A treble fishhook can be left in a dead bird as bait; the snake will swallow the hooks as it eats the bird and eventually die. Rats and snakes can get in through holes around doors and windows, so plug them up. To prevent chick mortality due to predators, thieves, and rain, coops or baskets can be used to house mother hens and chicks. They also allow for separate feed and water supplements, albeit because to the insufficient diet often provided in coops, some scavenging is still required.

Table 6: Predator attack modes and control methods

Predator	Attack mode	Control method
Hawk	Picks up stray birds and weaklings. Attacks birds so that head and toe marks are visible on back. Often plucks birds.	Hunt the hawk and keep chicks away from clear swoop areas.
Rat, mongoose	Usually take more than they eat, and stuff chicks in holes for later consumption.	If allowed, use rat poison.
Snake	Will swallow eggs and chicks	Use fishhooks.
Dog, cat	General destruction	Try to catch them. Cats can control rats but wild cats and dogs are a problem.
Fox, jackal	Will bite off the feathers over the back and between wings, eat the entrails and breast, and carry bird to den.	Roam in the early morning; kill for their young. Trapping is the best control.
Raccoon	Pulls off head and eats crop. Will carry birds off.	May be protected in some countries. A permit to destroy may be required.

Source: http://www.fao.org/3/y5169e/y5169e05.htm

Housing in Free-Range Systems

Under free-range systems, roomy, clean, and airy overnight housing should be provided. Houses can be permanent or moveable. If space allows, a movable chicken house may be ideal, and mobile folds or field units for laying hens can be installed to increase egg production. On the range, these mobile units can be rotated. Although shelter is less expensive and balanced feeds are not required, the birds are exposed to the sun and are susceptible to parasite infestation. On pasture, stocking density should be established based on soil type and pasture management strategy. Any existing structure, such as the farmer's outhouse, kitchen, or dwelling, can be converted into a night shelter for up to 20 free-range hens. There should be no more than three to four birds per square meter in a deep litter system. In areas where it rains a lot, the floor should be elevated with a large roof overhang, especially above the entry. A solid earthen platform or a raised bamboo platform can be used as the raised floor. The raised bamboo platform has the benefit of giving air beneath the chickens, which helps to keep them cool in hot weather and keeps them out of flood water during the monsoons. The building's walls can be composed of mud or bamboo, with bamboo slats for windows and doors. The house can also be self-contained, and it can be used in semi-intensive or intensive manufacturing systems.

Housing in Semi-Intensive and Intensive Systems

Planning

Complete confinement is only advisable where:

- there is good management;
- reproduction is spread equally over the year;
- land is scarce or inaccessible all year round;
- balanced rations are available;
- a supply of hybrid day-old chicks is available;
- labor is expensive;
- parasite and disease control are readily available; and
- the objective is commercial production.

The reasons for confinement are, in order of priority, to:

- reduce mortality due to predation in chicks under two months of age;
- achieve higher daily gain and better feed conversion in growers; and
- allow better supervision of production in laying hens.

The placement and building design of any constrained systems must be carefully studied. The grass or grazed area around the home should be mowed or grazed. The following requirements should be met by a good location:

- It should be easily accessible.
- There should be a reliable water supply.
- The ground should be well drained.
- It should be at a sufficient distance from residential areas (far enough to protect human health and close enough to provide security for the birds).
- It should be well away from woodland.

Existing facilities can be converted to provide housing, albeit planning clearance may be required. For example, a disused outhouse kitchen can be turned into a poultry house. In all conversions, meticulous planning should be used to make the most of the available space:

- A plan of the building should be drawn to scale.
- Use should be made of existing floors and walls, if suitable.
- Space requirements of the birds and manure disposal should be taken into consideration.
- A feasibility study should be carried out, taking into consideration future and requirements as well as the economics of converting the building.

Construction

The importance of the floor cannot be overstated. A well-drained concrete floor with a layer of heavy gravel or wire mesh embedded in it to keep rats away is excellent for a deep litter house. This style of flooring is usually quite expensive. Wood, bamboo, bricks, or huge flat stones (depending on what's available locally) can be utilized, but they're more difficult to clean. Clay floors are less expensive, but they require a fresh layer of clay to be applied either between flock batches or at least once a year. Raised floors are occasionally used in areas where construction materials are less expensive than deep litter, especially in humid climates where litter is scarce. To allow the droppings to collect under the home, they are composed of wire mesh, expanded metal, wooden slats, or split bamboo, and should be about one meter above the ground to allow for cleaning and ventilation. Higher storeys may cause the structure to become unstable. They are supported by pillars made of materials such as wood, bamboo, oil drums, and concrete blocks that are either rot-resistant or have stone or concrete footings. Rats can be kept out of houses with raised flooring on posts by using baffles. A metal collar, an upside-down tin can, or a metal band coiled around the post can all be used as baffles, but they must fit snugly to dissuade even the tiniest mouse. If the building is constructed sufficiently rat-proof, the roof and walls of the house can be formed of any affordable local material, such as bamboo slats, sorghum stalks, mud, wooden slats, and palm fronds. Walls should be thicker or insulated in colder areas, although thatch can be utilized in warmer climes, though it should be renewed periodically to avoid parasite and disease problems. To avoid tick and mite infestation and make cleaning easier, the inside of the walls should be as smooth as possible. Building walls that run the length of the building are not recommended since they restrict cross-flow ventilation. If the windows do not have shutters, the roof should be watertight and overhang the walls by one meter. Thatch, sheet metal, or tiles can be used for the roof. The cheapest option is usually thatch, which also provides good insulation. It will most likely need to be replaced every three years, or sooner if ticks get inside. To keep predators out, it should be interwoven with bamboo or wooden slats. Sheet metal is frequently too expensive, and it must be coated white or aluminum in hot countries to deflect sun heat. It is, however, easy to clean, which is a significant benefit in areas where ticks are a concern. Rain and rodents are kept out with a layer of plastic sheeting wedged between bamboo slats. Oil drums that have been flattened can be used at a lesser cost. Although sun- or oven-baked tiles are normally more expensive than thatch, they will endure far longer. The frame for a tiled roof must be stronger than for other materials due to their weight. The style of your windows is determined by the climate in your area. Chickens require more ventilation than humans, however they should be protected from the elements such as wind, dust, and rain. During storms, wood or bamboo hinged shutters or feed sack curtains can be used to cover window openings on the house's windward side. To limit the amount of moisture in the house in humid conditions, window design should take use of the wind direction as much as feasible. Wire mesh or expanded metal work well for window regions. Depending on available budgets and materials, wooden slats or bamboo can be used. The thicker the material, however, the less ventilation there will be. Metal, wood, or bamboo should be used for doors. Wire mesh could be used on the top half of the door. Doors should be robust enough to survive being opened and closed repeatedly throughout the year. When compared to flat or lean-to roofs, gabled roofs lower solar heat loading. A gabled roof's pitch, or "angle of rise," is critical for a variety of reasons. Traditional village thatched gabled roofs are often made of bush timber and pitched at a steep angle (more than 42 degrees from the horizontal), which helps the roof resist strong winds. In heavy winds, shallower pitched roofs are more likely to be blown off, especially when the pitch angle is 15° to 20°. Shallower pitched roofs have a smaller roof surface area, which lowers the cost of surfacing material; however, because they are more vulnerable to stormy winds, they require stronger support frameworks, resulting in a significantly higher overall roof cost. The best compromise between roof surfacing and roof support expenses is a 42° pitch. When there is a slight breeze and air movement across the shed at the height of the bird, the maximum width for an open-sided poultry house is 8 m. (26 ft). The end walls of the shed should be closed to increase the volume and velocity of airflow over the shed width. This forces the air to travel across the width of the shed, especially if the wind isn't already blowing from that direction. The use of a central ridge ventilation system is not suggested since it prevents airflow across the entire shed width. The air enters on the prevailing wind side, is drawn up in the center, and then exits at the crest, excluding the other half of the structure.

Nests: One nest should be supplied for every five hens to avoid overcrowding and eggs being placed on the floor. At least one square meter per 50 birds should be allowed if larger communal nests are employed. Nests can be made from baskets, pots, and cardboard boxes. A 25 cm base diameter, 18 cm high walls, and a 40 cm open top diameter are ideal for a basket or pot nest. Nests should be kept clean and secure in a shady, isolated location out of the sun. They should be lined with fresh litter. For bigger groups of hens, nest boxes for individual hens should be built in multiple groups. These are usually built of wood and should have a nest floor surface of about 0.1 m2 and measure approximately 30 cm on all sides.

Perches and roosts: Chickens love to sleep on perches at night. Each bird should have a perching space of 15 to 20 cm. During the day, birds below in the social pecking order can also use the perches. Each perch bar should have a cross-section of 2 to 3 cm. The length of these structures is determined by the number of birds to be kept. The perches should be set up in a frame, parallel to the wall and horizontally, with a sliding, removable platform called a "droppings board" about 20 cm below the perches to catch the dung droppings. The first perch bar should be 20 to 25 cm from the back wall, with following perch bars spaced 30 to 40 cm apart. The droppings board should reach 30 cm in front of the front perch bar and contact the back wall, allowing the birds to land from their flight from the floor before locating a perching location. To make cleaning of the droppings board easier, the perch bars should be about 20 cm above the droppings board and the droppings board should be 75 cm from the floor of the home. The usage of droppings boards helps to keep the floor clean because fowls deposit nearly half of their droppings at night. The manure can then be readily collected, dried, and stored in empty feed sacks for use as a great organic nitrogen fertilizer for plants. The space behind the droppings board is then excellent for a shared nest.

Providing Feed

Laying hens need regular access to food and water in both intensive and semi-intensive systems, and feeders should be evenly located throughout the chicken house. Birds scavenge during the day in the semi-intensive system, primarily for protein (from insects, worms, and larva), minerals (from stones, grits, and shells), and vitamins (from leafy greens, oil palm, and nuts), but energy supplements such as maize, sorghum, and millet are important for higher productivity and should be provided. Ingredients and feeding systems are covered more thoroughly in Chapter 3 on Feed Resources.

Feeders

A good feeder should be:

- durable enough to withstand frequent cleaning;
- stable enough not to be knocked over;
- of the correct height and depth;
- bird proof (such that birds cannot get into it or roost in it); and
- equipped with a lip to prevent birds from spooning feed out onto the floor with their beaks.

To prevent them from scraping contaminated litter into the feeders and to reduce feed waste, the height of the feed inside the feeder, which should never be more than one-third full, should be level with the backs of the birds. This is accomplished by altering the feeder's height. Feed should be provided at daybreak and around 14.00 hours (or more frequently if the birds empty the feeder), with all feed eaten before nightfall, to avoid spoiling and mould problems. To keep rats out, feeders made of wood, sheet metal, or bamboo should be hanging from the roof. The feeder's height should be adjustable. Supplementary vegetable matter should be given at beak level, either suspended from the ceiling or placed in a net or a floor-standing hopper with wire or slatted sides. It is not to be thrown on the ground. The linear distance of lip available to the birds is used to calculate feeder space. If the birds feed from both sides, this is either the circle of a round tube-feeder tray or twice the length of a trough. If troughs are utilized, each bird should have at least 10 cm of feeding space. There should be at least 4 cm of feeding space per bird when using circular feeders.

Table 7: Feed and feeder space requirements for 100 chickens

Age (weeks)	**Daily feed consumption (kg)**	**Suggested feeder depth (cm)**	**Feeder space (m)**
1 - 4	1.4 - 5.0	5	2.5
4 - 6	3.2 - 7.3	8	3.8
6 - 9	5.0 - 9.5	9	6.1
10 - 14	7.3 - 15.9	12.5	9.6
15 and above	9.1 - 11.4	15	12.7

Source: http://www.fao.org/3/y5169e/y5169e05.htm

Creep Feeders:

Creep feeders are used to allow baby chicks access to high-quality (rich in calorie and protein) feed while preventing entrance to larger birds (particularly the chicks' mothers) through a small doorway. The conical creep feeder (open at the top and bottom) is fashioned from 0.5 to 1 cm broad split bamboo strips bound at the joints with string or wire. The creep feeder's base can be 75 cm in diameter and 70 cm high, with a 20 cm wide access hole (reinforced to serve as a carrying handle) at the top. At the bottom, the spaces between the upright slats can be 2 to 3 cm, and at the top, they can be around 1 cm. The bamboo strips' flexibility allows the size of the entryways to be increased as the chicks grow larger. If the chicks are hesitant to leave their mothers' side, a more densely constructed weave can prevent them from exiting the cone once they've been placed inside through the top entrance. A covering of old engine oil helps protect the bamboo from insect assault. By building full immunity, better nutrition for young stock enhances their immune response to illness challenge and vaccine response. The use of creep feeders could lower chick death rates; however, it did not boost growth rate. The development and survival rate of chicks rose when the home trash was supplemented with protein and fed in the creep feeder. If their chicks are fed in a creep feeder after hatching, annual egg production can be doubled due to the increased laying time available to the hens. If the gaps in the creep feeder are adjusted, it can also be used for growers over eight weeks, and if given less than the full ration, they will learn to compete for food with other chickens.

Table 8: The effect of creep feeding on flock egg production

Intervention	**Period**					
	June	**July**	**Aug.**	**Sept.**	**Oct.**	**Nov.**
Creep only	31.5	28.7	27.3	21.8	21.4	33.0
Creep + Low supplement	21.2	18.8	22.9	26.9	30.7	31.1
Creep + High supplement	24.3	24.5	32.5	34.1	27.4	31.1

Source: http://www.fao.org/3/y5169e/y5169e05.htm

Providing Water

Providing clean water is a priority that is frequently overlooked. The amount of water, the correct type of equipment, and the location are all critical factors to consider. Water consumption rates for hot, dry circumstances are shown in the table below, which can be decreased for temperate climates. The only specific action to be taken in countries with abundant of water, such as Bangladesh, Vietnam, Indonesia, Gambia, Sierra Leone, Zaire, and Uganda, is to maintain its cleanliness. Water fetching and carrying is a critical chore in other places, particularly in the Sahel and other drought-prone areas, and is generally entrusted to women and children. Because brooding chickens are kept inside during the rainy season to avoid contamination with mud and filth, clean water and feed must be provided in the home. A tin can inverted into a soup dish or the bottom of a larger tin can is the most basic equipment. About 2 cm from the open end of the tin can, a hole is punched. The can is filled with water, the plate is placed on top, and both are immediately inverted. The water level in the plate is controlled by the position of the punched hole and the vacuum in the tin can. Tin can waterers are effective, although they rust quickly, especially in the humid tropics. Adult birds can be watered with a clay container or gourd with holes around the edges lowered into the ground for stability. Local potters can usually manufacture any size clay pot that you want. They deliver water that is half a degree Celsius colder than other waterers because they are permeable, and heat loss through evaporation keeps the water cooler. This also implies that the pots will lose a significant amount of water over time, especially in extremely hot and dry climates, necessitating the use of glazing to make them more impermeable. If continuous drinking troughs are employed, each bird should have at least 5 cm of trough. Alternatively, one cup or nipple drinker for every ten birds might be provided.

Table 9: Minimum water and watering space requirements for 100 birds in hot dry conditions

Age (weeks)	Daily consumption (litres)	Water space (m)
0 - 1	3	0.7
2 - 4	10	1.0
4 - 9	20	1.5
9 or more	25	2.0
Layer	50	2.5

Source: http://www.fao.org/3/y5169e/y5169e05.htm

Baby Chick Management

Warmth and dryness are essential for baby chicks. They must keep the nest clean, which they share with the mother hen at night. The nest location should be kept warm in colder climes (below 20 °C at night) by filling it with straw and placing it near a stove or fireplace. The chicks should stay with the mother hen during nine to ten weeks, learning how to scavenge and avoid predators and other threats from her example. To augment scavenging, clean drinking water and new feed in a clean container should be offered. For further information on feeding procedures." The weight and growth of chicks, as well as their mortality rates, are all linked. In a study in which young chicks were given supplemental protein feed via a creep feeder (Roberts et al., 1994), it was discovered that supplementary protein feed had a significant impact on survival and growth rates. Chicks that were separated from their mother hens during the day from the age of three to ten weeks and fed ad libitum chicken starter mash had a mortality rate of 20% and a body weight of 319 g at ten weeks, compared to a mortality rate of 30% and a body weight of 242 g for the control group that stayed with the mother hens. More information can be found in the section above titled "Creep Feeding." As a result, the following would be a good method for raising chicks:

- The chicks should be kept confined for the first several weeks of their lives and fed a well-balanced diet.
- It is necessary to adhere to a vaccination schedule.
- During the remaining growing period, sufficient supplemental feed should be provided to allow the hens to develop according to their genetic potential.
- During the first four to eight weeks of life, feed supplements and protection should be given to naturally brooded hens.

The supplementary feed's composition will be determined by the scavenge-able feed available, but a cafeteria-style free-choice feeding system with a protein concentrate, energy concentrate, and calcium mineral in each of three containers may be the best option. Natural brooded chicks, whose primary source of food is scavenging under free-range settings, have a significant mortality rate that sometimes surpasses 50% until they are eight weeks old. Predators were responsible for up to 88 percent of mortality, and colored birds fared better than white birds in terms of survival. Low offtake from scavenging poultry flocks is due to a high mortality rate and the enormous quantity of eggs required for hatching. Confining and vaccinating the chicks during the raising stage is an effective technique of lowering mortality (a expensive loss). However, this is more expensive, with feed expenses in particular driving up production prices. Many poultry development programs in Bangladesh have utilized a system that confines chicks during the first eight weeks of their lives for the previous ten years. They are given around 2 kg of balanced diet each and then kept in semi-scavenging circumstances. Because of their bigger body weight and more efficient vaccine inoculation, they are less vulnerable to predator attacks and more resistant to infections at eight weeks of age (due to their better nutrient intake).

Hygiene

Manure Management

Manure control must be taken care of regardless of the form of confinement. Adult birds create 500 g of fresh manure per kilogram of body weight per year (70 percent moisture content). Manure should be dried to around 10% to 12% moisture content before storage to preserve its fertilizing potential. This will ensure that the maximum amount of nitrogen is retained for fertilizer value. The most volatile component of manure is nitrogen in the form of urea, which is lost as ammonia if the moisture concentration is too high in the stored material. When the moisture level of stored manure is too high, ammonia, carbon dioxide, hydrogen sulphide, and methane are released, all of which can have catastrophic physiological effects on humans. Some of these elements are also greenhouse gases, which contribute to the global warming of the atmosphere. Poultry manure is an excellent organic fertilizer, animal and fish feed, and a raw material to produce methane gas in biogas plants for cooking fuel.

Other Hygiene Management Measures

Diseases and pests are less likely to spread when there is adequate ventilation. Perches or loosely plaited bamboo mats (such as those used for sifting) placed on the floor in overnight dwellings can help keep them dry. If the birds are kept inside, the floor should be swept at least once a day. To disrupt the reproduction cycle of the common housefly, clean an exterior chicken house once a week. The entire breeding cycle, from fly egg to adult housefly hatching, takes roughly seven days. Spreading wood ash and sand on the floor will keep lice at bay. Mothballs (naphthalene) crushed with ash can also be applied to the birds' feathers or wings, or placed where the chickens normally dust bathe. If the hens are already afflicted with mites, the home can be fumigated with a kerosene-soaked rag (while the chickens are outside). Birds carry lice, so dust baths using naphthalene powder in the ash will be more effective than dust alone. It is recommended that hens and ducks not be kept together. Wet flooring result, which can lead to infections like Fowl Cholera. Ducks are likewise far more resistant to Newcastle Disease than chickens, and are hence frequently carriers of the virus. To avoid cross-infections and bullying injuries, chicken adults and young stock should be housed separately.

Management of Free-Range Poultry

Poultry that is allowed to roam freely is frequently an issue. They transgress into neighboring fields and gardens, and they are continuously preyed upon. Because of the cost of feed and fence, confinement is rarely practicable, and surveillance is only viable when the elderly or young members of the household have time to assist. In many circumstances, fencing vegetable patches is the best option. More cocks in the hamlet might minimize the number of chickens moving around because each flock's cocks and hens would stick to their own territory. Cocks occupy an eight-to-ten-house territory, whereas hens occupy two or three. The discrepancy between the quantity of food acquired through scavenging and the overall food need for maximum production should be balanced with nutrients supplied by supplementary feed in the free-range system. To build a well-balanced supplement, you'll need to know the scavenger feed resource base (SFRB) and the crop content composition. If this is unknown, it is recommended that the fowls have access to three containers (or three compartments of a bamboo stem feeder) of components containing a protein concentration, a carbohydrate supply (for energy), and a mineral source (using a free-choice cafeteria system) (mainly for calcium carbonate for egg shell formation for the hen). To complement the day's scavenging, poultry should have free access to this cafeteria system for two to three hours in the evening. From the standpoint of feed resources, this advice is only economically viable (sustainable) if the extra feed consumption per egg produced is 150 to 180 g or less. Over-consumption of supplements is only justified if they are less expensive than the commercial feed used in intensive poultry farming. Supplements are typically recommended in the range of 50 to 80 g/bird/day, making it a realistic option. Seasonal changes in the SFRB have a significant impact on productivity. During the dry season, scavenged feed (such as grass shoots, seeds, worms, insects, and snails) from gardens, crops, and wasteland ceases, while the quantity and quality of home cooking trash decreases. To maintain an optimal level of production, the feed supplement should be modified seasonally, or the chicken population could be adjusted to the amount of SFRB and feed supplement. In confinement, hens fed a healthy diet convert food weight to egg weight at a rate of 2.8 kg of feed per kilogram of egg weight. Changes in husbandry may be enough to boost the output of scavenging village hens without requiring any additional inputs. A stocking rate of 120 to 180 birds/ha in planted orchards will clean up windfalls while also nourishing the trees. In this example, the assumption is that the amount of fertilizer generated per hectare of 500 g of fresh (70 percent moisture) weight of manure is produced per kilogram of live weight per year for 150 hens (weighing two kilograms a piece). This equates to 330 g of dried manure per hen per year (dried to a ten percent moisture content), or 49.5 kg of dry manure per year for the 150 hens. This is equivalent to 13 percent ammonium nitrate, 8.6 percent superphosphate, and 2.9 percent potash (potassium) salts in terms of fertilizer value. As a result, the 150 hens will generate 6.4 kg of ammonium nitrate, 4.3 kg of superphosphate, and 1.4 kilogram of potash salts per hectare every year.

Planning Flock Production and Size

Birds are used in the production of meat and eggs. The number of chickens in the flock is the most essential factor in both meat and egg output. As eggs hatch and hens are sold or eaten, the size of the flock fluctuates. The most common cause of flock reduction is death, especially among chicks. Disease is the leading cause of death, especially during the rainy season and humid seasons on either side of the equator. Predators in cropped fields contribute to reduced flock sizes throughout the summer and wet season. Local birds lay three to four clutches of 12 to 15 eggs per year on average, with more eggs laid during crop harvest because there is more food available. Keeping the flock number constant in most conventional farming systems necessitates eight to ten eggs for reproduction, leaving an average of 35 to 40 eggs each layer for sale or consumption. The extra eggs can be sold or consumed because the quantity of eggs required for replacement may decrease with improved management. Most of the egg laying occurs between daybreak and mid-morning. Nest placement should not be moved during the laying months, since this may disrupt the laying routine. The sale of eggs and live birds provides cash to local flocks. A flock of 15 local hens laying 30 eggs per hen per year (plus one local cock) would generate 450 eggs in a year. 120 eggs may be incubated by broody chickens (in ten clutches of 12 eggs each), with 100 chicks hatching, and 30 eggs may be cracked and consumed in the home, leaving 300 eggs for sale. Thirty of the 100-day-old chicks may achieve maturity (with 70 percent rearing losses), yielding 15 cockerels and 15 pullets. The 15 pullets will replace the older hens, of which ten will survive the cull hen sale, and one young cockerel will replace the old cock. As a result, the annual income from the flock can be computed as follows:

300 eggs + 10 old hens + 1 old cock + 14 cockerels = income

Culling is crucial for increased output, and productive birds should be carefully picked. For the sake of simplicity, no adult mortality is assumed in the preceding case.

SMALL-SCALE BEEKEEPING

Introduction

Beekeeping is a good example of a small-scale agricultural development activity. Since World War II, the number of beekeepers in the United States has progressively decreased, owing in part to land development, which has destroyed many flowering plants from which bees collect nectar to manufacture honey. Nearly 100,000 people in the United States, however, raise bees for fun or profit. Major metropolitan regions such as New York, Detroit, and Los Angeles have established beekeeping laws in the last decade, and membership in metropolitan beekeeper's societies has increased to match. Even well-known luxury hotels, such as the Waldorf Astoria and the Fairmont San Francisco, have gotten in on the act, constructing hives with hundreds of thousands of bees to offer pollinators for their restaurants' hotel gardens and to help the environment. Beekeeping is a suitable back-to-nature pastime for city residents because of the modest scale of bee colonies. You simply need a square yard [measurement] to have a beehive in terms of space. Bees have several advantages over other creatures, including the fact that they do not require as much care. Most animals require you to purchase all their food and feed them daily. You must also let them out in the morning and return them to their beds at night (unless you live on a farm). Bees, on the other hand, can take care of themselves. They will fly to the source of their food and eat it. They'll also let themselves out and go to bed by themselves. As a result, you can leave them for quite some time. When you look at them, you might find that you have an excess of honey to take.

Start Small-Scale Beekeeping

Because most parts of the world have a bee-human link, any beekeeping development effort should aim to bring new and more efficient ways. The bee resource is already available; the goal is to make better use of it. All the materials needed to start a beekeeping business can be made locally. Local tinsmiths, tailors, carpenters, and basket craftsmen can build smokers, protective garments, veils, and hives. As a result, a beekeeping project can provide employment and income for these individuals. From the start, a little beekeeping project can be profitable. It is simple for a beekeeper to increase the quantities of hives after a project is started and competence is obtained. It is not required to rely on external resources or inputs to do this. Bee's forage much beyond the limited amount of ground on which the hives are located, feeding themselves from the area's existing nectar and pollen resources. Beekeeping is a family-run business. Although dealing with bees is a simple task for women, it is generally considered a man's task in most cultural settings. Women are often involved in the preparation of honey for market and in the actual commercialization, whereas men labor directly with the bees. Honey is typically viewed as a cash crop rather than a product for personal consumption by small producers. Honey has a significant monetary worth in comparison to its weight and volume. It's essentially a non-perishable product that's cheap and easy to carry when properly preserved. Honey is an appealing crop for small-scale and often isolated producers because of these properties. While most locations have a ready market for honey, the same cannot always be said for beeswax. It may be required to build a market for the wax in some locations. Beeswax is a non-perishable, readily kept substance. Local artisans and craftsmen such as lost wax metal (typically brass) casters, wax printers and batikers of textiles, tanners and leather workers, and candle makers employ it in some locations. Beeswax is also commonly used in the production of wood polishes. Pollen and propolis have markets in areas of the world where the beekeeping sector is well-developed (tree resin gathered by bees for use in the hive). While these may be viable items for a beekeeping business, they are not appropriate for beginners. Pollen production is challenging, and in most places, there are few local marketing avenues for these items. Bees themselves are a byproduct of beekeeping. When beekeeping becomes popular, a market for bees emerges. Some beekeepers can provide bees to those who are interested in starting a beekeeping business. While bee products such as honey and beeswax have ready international markets, any development effort should prioritize local markets. A small project does not produce enough to warrant looking for an international market. Creating a local market for bee products protects local producers from global price fluctuations while also providing a market for small producers.

Planning For beehive Project

The first step in designing a beekeeping operation is to learn about the relationship between bees and humans in your area. Talk to those who work with bees in your region. When they're working with bees, accompany them. If you've never worked with bees before, working with local bee-killers, bee-havers, or beekeepers can teach you a lot. You will be able to make more credible suggestions for improvement if you understand how they work, and it will be easier to adapt a suitable beekeeping technology to the region if you understand how they work. Also, before committing to beekeeping, you should experience a few stinging occurrences. Beekeeping necessitates the use of bee stings. They must be dealt with by a beekeeper. Ideas for implementing improved approaches should be formulated once you have a better understanding of the local bee-human relationship. With whom should I collaborate? What is the best equipment to use? Where should hive goods be sold? If you're new to beekeeping, it's generally best to start with just one or two people in the neighborhood. Your efforts will be amplified if you choose farmers who are well-liked by their peers and have strong community connections. Keeping bees for yourself and employing methods that aren't often used in the area is a good start. Sooner or later, word will get around, and you'll be chatting about beekeeping with your friends and neighbors. Always start with at least two hives when beekeeping. This allows you to compare the growth of multiple hives and, more crucially, it permits the project to continue if one colony dies away. In addition, administration directed at an apiary rather than individual hives might be stressful. Set reasonable goals while planning a project. A successful modest endeavor is more meaningful than a massive project that is attempted but fails. The pace of change is gradual. It must begin with a concept. In some cases, presenting a concept successfully is a realistic expectation for introducing improved approaches to the bee-human relationship. The equipment to be employed in a project is determined by the conditions on the ground. When deciding the type or types of hive equipment to use, you should consider the availability of required inputs as well as the professional assistance available. Finding someone who can create beekeeping equipment in your area and getting it made can be a success in and of itself. Coordination of putting the equipment together can take a lot of patience. In most regions, hive products can be sold through local marketing outlets. Check to see if someone is already utilizing honey or beeswax. They frequently express a desire for a consistent supply of a high-quality goods. Local bakers and confectionery producers are a possible market if they aren't currently using honey. Check with anyone who might be able to help you find a market for your beeswax.

Beekeeping is a project that fits in nicely with other agricultural and rural development initiatives. Beekeeping ventures can also be implemented as part of regional development programs. Certain crops cultivated in such schemes can produce honey for the beekeeper while also benefiting from bee pollination. Small farmers who plant these crops or keep their bees nearby may be able to supplement their income through beekeeping. Beekeeping projects can be set up at vocational and agricultural training centers, as well as rural teacher training institutes. The trainees can have a multiplier effect when it comes to introducing beekeeping to villages, and the centers themselves provide good demonstration places. Beekeeping is also a fantastic project for a school or youth club. Farmers of the future are students and members of youth groups. They are receptive to new ideas and procedures and can assist their parents understand them. A school demonstration apiary not only gives a good educational opportunity, but it can also provide some cash. Beekeeping and honey sales are also promoted through exhibits of bees, equipment, and hive goods at area fairs and gatherings. Demonstrations of bees in action can help dispel many common misconceptions about bees. Making an artificial swarm and then re-hiving it is a very attention-getting demonstration. A glass-sided observation hive can pique people's interest in bees. It provides an opportunity to observe bees at work in their hives, making it an excellent educational tool. Because such a small hive generally necessitates close attention and care, it provides several possibilities for teaching the colony's management demands and practices. Beekeeping is also a good fit for cooperatives. As part of their activities, several cooperatives operate beekeeping projects. Beekeeping is the sole activity in some coops. These apiaries provide critical inputs, technical assistance, and markets for honey and beeswax. Beekeeping coops have shown to be highly successful in several circumstances.

Types of Bees

Bees come in a variety of shapes and sizes. Although most bees are solitary, a few are sociable. These folks live in colonies with a division of labor among themselves. All bees are significant pollinators because of their habit of visiting flowers. Although all bees collect nectar and pollen from blooms, only a small percentage of social bees store the nectar as honey. There are even fewer species of bees that store honey in sufficient quantities to make the labor of extracting honey profitable. Although some "stingless" bees in tropical areas are robbed of their honey, real honey bees of the species Apis are the main providers of honey and other hive products. The western hive bee, Apis mellifera, has been brought to most parts of the world for use in beekeeping.

Stingless bees: Some stingless bee species, particularly Trigona and Melipona, are robbed of their honey in tropical areas. The nests of all these bees are built inside cavities. Even though they do not sting, these bees defend their hive by biting intruders. Some bites release irritants in addition to the bite. Stingless bees' brood comb is one cell thick and usually horizontal. Honey is stored in thimble-sized wax honey pots surrounding the brood area of the nest by these bees. These stingless bees are housed in gourds, clay pots, and hollowed logs in some locations. Honey is obtained by removing the honey pots from the nest cavity. Because the output is so low, marketing it is only worthwhile on a local basis. Locals cherish such honey for its curative properties. Central American Mayas devised a mechanism for keeping one species of Melipona in log hives. Honey played a major role in their religious rites, and they held specific festivals and rituals to ensure a plentiful harvest of honey. This is the first and only time in the world that a non-Apis species of bee has established a bee culture. In some parts of Central America, meliponiculture is still practiced. These bees provide maximum yields of 10-12 kilogram per year. Improved methods for retaining stingless bees are restricted in their efficiency for boosting production since they require a different type of nest building for honey storage and brood. Because of this, and because of the low yields, maintaining these bees is only economically viable for home honey consumption.

Apis (The true honey bee): The Apis bee genus has four species, three of which are native to Asia and one to the Euro-African region. Although there are differences in size and color, they all have a similar appearance. All of them make two-cell thick vertical combs. Asia is home to the huge or rock honey bee (Apis dorsata) and the small honey bee (Apis florea). Both bees create an uncovered single-comb nest. Nests are frequently observed hanging from tree branches, roofs, and ceilings. To keep the nest conditions under control, the adult bees weave curtains around it. Brood and honey are both stored in the same comb, with brood in the lower area and honey in the upper. Methods for detaching the honey region of the comb and reattaching the brood area have been devised in some areas, and so bee keeping is done with these bees. The yields are frequently high enough (particularly with A. dorsata) to make the farmer's efforts profitable. There are no known ways for raising either of these bees that are better than what is currently used. Both species' behavior is erratic, and they will not live in a hive. The enormous honey bee is particularly protective of its hive. As a result, there is limited room for improvement in either species' management, while there is typically room for improvement in honey quality by taking more care in processing. Other Apis species (mellifera and cerana) usually construct multi-comb nests in enclosed cavities. These bees can be kept in hives, and ways have been developed to make better use of their capabilities. There is promise for beekeeping development with these two species. Western Asia, Europe, and Africa are home to the western hive bee (Apis mellifera). There are at least twenty different sub-species or "races" documented across this bee's range, which are largely separated into European and African groupings. Several races of this bee are thought to be particularly good for beekeeping. Most sections of the planet, including the Americas, Australia, and Asia, have been colonized by European races of the western hive bee. This bee has been thoroughly researched from both a biological and a beekeeping standpoint. Desirable breeds build big colonies and generate large surpluses of honey when conditions are favorable. Under ideal conditions, yields of 100 kg/

year or higher are feasible. The western honey bee has a lot of potential in terms of beekeeping. Aside from high honey production, this bee's ability to thrive in a wide range of environments, as well as its availability due to past introductions or local populations, have made it a popular beekeeper. The focus of this guidebook is also on the western hive bee for these reasons. Asia is home to the eastern or Indian hive bee (Apis cerana, originally A. indica). Because it can be easily hived in man-made containers, beekeeping developed in several parts of Asia with this bee. Some places produce up to 15-20 kg of honey per year, although the average is significantly lower. The eastern hive bee has a lot of variety across its range, and little work has been done to select more attractive strains from a beekeeping standpoint. Beekeeping techniques for this bee are like those for the western hive bee, except the hives are smaller. In its native habitat, the development of small-scale beekeeping with the eastern hive bee warrants study. Although yields are smaller than those of the western hive bee, this bee has the advantage of being well acclimated to the local environment. It is more resistant to some of the disease and pest issues that are prevalent in the area. As a result, it is better able to thrive under the minimal management circumstances that are common in small-farmer beekeeping. In Asia, there has been a push to replace eastern honey bees with European races of the western honey bee in recent years. This has only worked in temperate climates and for large-scale, capital-intensive operations where the capability to address disease and parasite problems in European breeds is available. Consider the expected beekeeping circumstances while selecting a species for a small-scale development effort in this area. Importing bees is frequently laden with difficulties. Importing bees for a small-scale enterprise can lead to individuals being psychologically reliant on outside help. They are unaware that a local bee resource exists, and they are unmotivated to take advantage of it. Imported bees are frequently unadapted to the environments in which they are released. This is especially evident when temperate European races are introduced into tropical areas. Importing bees also increases the possibility of exotic bee illnesses and parasites being introduced. With the introduction of Varroa, a honey bee mite parasite, tragic incidents have happened in Europe and South America in recent years. The most well-known incident of uncontrolled bee immigration occurred in Brazil in 1956. In many temperate South American locales, a beekeeping business based on European races of the western bee was well-established. To build an industry in some of Brazil's more tropical regions, African bees were introduced. Some of these bees escaped and established themselves. They've continued to spread across the tropical lowlands, and in many areas, they've displaced European bees. The introduction of African bees to tropical America has wreaked havoc on the beekeeping business. The African bee is known for its vigilance and unpredictable behavior. From the perspective of a beekeeper, these are undesirable traits. Nearby people and farm animals have been stung to death in certain circumstances. It's been tough for beekeepers to adjust to the increased frequency of stinging episodes and the complexity of managing this bee. When bees become "Africanized" through interbreeding with wild colonies, it is frequently necessary to relocate hives away from populated areas. The African bee, on the other hand, is often more suited to low-management beekeeping than the European bee. The natural colonies of the African bee provide an inexpensive source of bees for small-scale farmers who want to start beekeeping. Because the African bee is designed to live in tropical climates, such colonies are prevalent in the locations where it is found.

Bee Traits Desirable for Beekeeping

- High honey production
- Gentleness
- Low tendency to swarm
- Low tendency to abscond
- Calm on combs when colony is worked
- Disease resistant
- Little use of propolis
- Little brood-rearing during dearth periods to conserve stores

Basic Starter Equipment

- To begin, you may get by with simply a bee veil and helmet, especially if money is an issue.
- Get a pair of goatskin bee gloves that are lengthy and soft. Even if you won't be wearing gloves all the time, having them on hand is a good idea.
- From time to time, you'll require a bee smoker. A hive that is becoming agitated can be calmed down with a little smoke.
- Get a couple of hive tools to help you separate frames and get them started outside the hive. Purchase two because they're easy to lose in the grass.

- Get yourself a frame puller. Nothing compares to it when it comes to handling a frame and thoroughly inspecting it.
- Make sure you have enough hive bodies and supers to accommodate all your frames and bees. You can use deep hive bodies for the brood nest and medium hive bodies for the honey supers, or medium hive bodies for everything. When full, mediums are easier to lift, and this arrangement makes it simpler to find comb. One disadvantage of using all medium-sized boxes is that finding the queen can be more difficult. If you're lucky, you'll be able to locate a local provider for frames and hive bodies, as shipping costs would kill you if you must send them. Buying them dismantled and gluing, nailing, and painting them yourself saves a lot of money.
- Begin with two or three three- or four-pound packages of bees. Because bees vary so greatly, one package is insufficient. This is a mistake I made, so don't make it.
- If you have the funds, invest in a ventilated bee jacket or bee suit.
- When things don't go as planned, a local mentor or bee club can offer support and encouragement.
- Choose "small-cell" beekeeping from the start. There's more on that later.
- Purchasing a few packages of bees is only the beginning. You should learn how to raise your own queens and locally adapted stock in the long run. It's not a sustainable strategy to rely on a few warm states to provide all the country's package bees. I'd like to see more northern beekeepers selling northern-adapted bees, and there are some encouraging signs in that regard (like a Northern Queen Breeders Association, for instance). To become as self-sufficient as possible, learn how to execute your own hive splits, the "Miller Method" of queen-rearing, and so on.

Climate and Beekeeping

To comprehend the relationship between climate and beekeeping, it is necessary to first comprehend two notions concerning bees and their surroundings. The nectar flow and the honey flow are the two types of flow. Although many beekeepers refer to them as the same thing, they are two distinct but related concepts.

The nectar Flow: The flow of nectar is entirely controlled by plants. It relates to the quantity (number of dissolved sugars) as well as the quality (quantity of dissolved sugars) of the nectar released by the plant. The amount of nectar flowing in each location at any given moment is determined by the plant species present and the weather conditions affecting those plants. The flora of an area, and hence the potential nectar flow, is determined by climatic and edaphic (soil) conditions. Rainfall, temperature, and sunlight all have an impact on the plants, which influences the actual nectar flow. Some plant species produce very little nectar, if any at all, while others produce a large amount. The quality of nectar, or the amount of sugar in it, varies by plant species. Quality is also influenced by the weather. High rainfall encourages nectar secretion, however nectar with low sugar content is common. For most plant species, enough rainfall before to flowering and dry, sunny conditions during the flowering period promote optimal nectar flow. Year to year, the time and relative amount of rainy and dry, sunny periods varies, making the nectar flow very unpredictable. Weather patterns have a lesser impact on some plant species than on others. Every year, good nectar flows can be counted on from these plants. Other plants are extremely sensitive to changes in the weather. These plants may produce excellent flows in some years while producing none in others.

The honey flow: Honey flow is determined by the connection between bees and plants. It is the honey bee colony's usage of the nectar flow. While the beekeeper has little control over the nectar supply, colony management is critical for excellent honey production. To maximize the honey flow, strong colonies are required during the peak nectar flow. The honey flow is also affected by the weather. A healthy honey flow requires suitable flying conditions for foragers during a good nectar flow. In terms of prospective honey flows, deciduous woodland areas in the wet/dry tropics are the best places on the planet. The protracted dry season in these places allows bee colonies to build up their strength to peak population levels, allowing them to take advantage of the maximum nectar flow. These places have a lot of melliferous (bee-attracting) species in their flora. After the wet season, the dry, sunny period supports a good nectar flow and provides ideal foraging conditions. Large apiaries of up to 100 colonies can be supported in these regions, with optimum yields of up to 150 kg/colony/year. Areas that are consistently chilly, overcast, or wet are unsuitable for beekeeping. The nectar is frequently of poor quality, and the bees have little opportunity to feed in these areas due to the lack of excellent weather. Of course, even if rich bee pasture is present for brief periods of the year, big desert locations are not suitable for permanent beekeeping. Bees aren't attracted to all plants. Within any major climate zone, the melliferous flora differs greatly. Natural elements have an impact on the environment, and land use patterns and agricultural practices can occasionally alter a region's appropriateness for beekeeping. If the imported crop is a poor resource for bees, cutting down vast amounts of suitable bee forage and dedicating these areas to monoculture can destroy a good bee region. In contrast, if marginal melliferous plants are replaced with good nectar- and pollen-producing plants, a bee pasture can be enhanced. Although substantial plantings exclusively for the goal of increasing bee pasture are rarely economically possible, an area's bee pasture can be improved by selecting good melliferous plants for other primary objectives such as reforestation, windbreaks, cover crops, fuel crops, or livestock feed crops. Under a variety of nectar and honey flow circumstances, beekeeping can be profitable. These factors are more important in deciding the size and type of a profitable beekeeping enterprise than they are in determining profitability. There are numerous sites that would be unsuitable for a large-scale beekeeping operation but would be ideal for a small-scale initiative. The importance of bee flora should not be overemphasized when establishing a small-scale beekeeping project. Most of these projects aim to improve an existing bee-human interaction by introducing new ways. It is significantly more feasible to presume that a bee pasture exists if a bee-human interaction already exists and the bees are producing honey. It's a waste of time and money to focus on the region's bee flora when starting a small-scale enterprise. The identification of bee flora is an important aspect of a beekeeper's experience-based expertise. It is not necessary to conduct a separate investigation. When there are beekeepers to watch it, it will be learned. Making beekeepers is the first step.

FISH FARMING

Overview

In many regions of the world, fish has been a significant element of the population's food for ages. Because of advances in technology, such as more powerful engines and sonar equipment, fish catches have increased dramatically during the last century. Overfishing resulted, resulting in a global decline in wild fish stocks. As a result, the increase in fish captures came to a halt about 20 years ago. As a result, the necessity to enhance fish production through farming has become critical. Aquaculture refers to the cultivation of aquatic animals and plants in freshwater, brackish water, and saltwater. Aquaculture shares the same goal as agriculture in that it aims to boost food production beyond what would be generated naturally. The live-weight volume of global aquaculture production more than from 34 Mt to 112 Mt between 1997 to 2017, (Naylor, 2021). Fish farming can be linked with agricultural, animal husbandry, and irrigation activities to maximize the use of local resources and, as a result, increase productivity and net profits. This method is known as "integrated fish farming." The following are the most significant benefits of fish farming:

- Fish is a good source of high-quality animal protein for humans.
- Fish farming may easily be integrated into an existing farm to generate additional revenue and better water management.
- Fish growth in ponds can be regulated since farmers choose the fish species they want to produce.
- The fish that are generated in a pond belong to the owner; they are safe and can be harvested whenever they want. In untamed seas, fish are freely available to everybody, making an individual's part of the common catch unpredictable.
- In a pond, the fish are frequently close by.
- Effective land use: Marginal land, such as land that is too poor or too expensive to drain for agriculture, can be successfully committed to fish farming if properly prepared.

Methods of Fish Farming

Fish farming can range from small backyard ponds to large-scale industrial operations. Input levels can be used to describe farming systems.

- **Extensive:** Economic and labor inputs are often low in large-scale fish farming. Natural food production is quite vital, and the system's output is quite low. Fertilizer can help boost fertility and, as a result, fish production.
- **Semi-intensive:** Semi-intensive fish farming necessitates a moderate number of inputs, and the use of fertilizer and/or supplemental feeding boosts fish productivity. Higher labor and feed expenses are a result, but higher fish yields usually make up for it.
- **Intensive:** Intensive fish farming entails a high level of inputs and many fish in the ponds. Supplemental feed is given to the

fish, with natural food production playing a small part. High fish stocking densities can generate challenging management issues in this system (increased susceptibility to diseases and dissolved oxygen shortage). Due to the high production expenses, a high market price must be obtained for the fish farm to be economically viable.

Pond Culture

Ponds are where most freshwater fish are reared. The pond is filled with water from a lake, river, well, or other natural source. The water may either run through the pond once and then be drained, or it may be partially refilled so that a portion of the total water in the system is retained. Water lost through evaporation and seepage is only replaced in pond systems that produce the most fish. In the tropics, water flow reduces the productivity of pond systems. The size of a fish farming pond can range from a few hundred square meters to several hectares (ha). Small ponds are utilized for spawning and the creation of juvenile fish, whereas larger ponds are used for the grow-out period. Most growers dislike production ponds larger than 10 hectares because they are difficult to manage. The ponds depicted here are merely examples. The type of pond a farmer constructs is mostly determined by local resources, equipment, and conditions. Ponds are frequently found on gently sloping soil. They're rectangular or square in shape, have well-finished dikes, and don't collect run-off from the nearby watershed. It's critical that there's enough water to fill all the ponds in a suitable amount of time and keep the same pond water level. When it's time to harvest the fish, you should be able to completely empty the pond. Side slopes should be 2:1 or 3:1 (per meter of height requires 2 or 3 meters of horizontal distance), allowing simple access to the pond while reducing erosion concerns. To avoid fish theft, situate the pond as close as feasible to your home. Another approach to keep thieves out of your fish pond is to submerge bamboo poles or branches in the water, which prevents netting and rod-and-line fishing. Aside from deterring theft, the poles and branches provide additional natural food for the fish. Periphyton-based fish aquaculture is the name for this method.

Table 10: Characteristics of a good pond for fish farming

Character	Description
Location	Take advantage of existing land contours by choosing land with a gradual slope.
Construction	Ponds can be sunk into the ground or raised above or below the original ground level. To minimize erosion and seepage, the slopes and bottom should be tightly packed during construction. A minimum of 25% clay should be present in the soil. The dikes should be free of rocks, grass, branches, and other unpleasant things.
Pond depth	Depth should be 0.5-1.0 m at shallow end, sloping to 1.5-2.0 m at the drain end.
Configuration	Best shape for ponds is rectangular or square.
Side slopes	Construct ponds with 2:1 or 3:1 slope on all sides.
Drain	Take advantage of existing land contours by choosing land with a gradual slope.
Inflow lines	Inflow lines should be of sufficient capacity to fill each pond within 3 days. If surface water is used, the incoming water should be filtered to remove undesirable plants or animals.
Total water volume	Sufficient water should be available to fill all ponds on the farm within a few weeks and to keep them full throughout the growing season.
Dikes	Dikes should be sufficiently wide enough to allow mowing. Dike roads should be made of gravel. Grass should be planted on all dikes.
Orientation	Locate ponds carefully to take advantage of water mixing by the wind. In areas where wind causes extensive wave erosion of dikes, place long sides of pond at right angles to the prevailing wind. Use hedge or tree wind breaks where necessary.

Source: Carballo et al., (2008).

Different Pond Types

Depending on the site, there are two different types of fish ponds to choose from:

Diversion ponds: Diversion ponds are constructed by bringing water from another source to the pond. Below are the different types of diversion ponds:

- **Embankment ponds:** The embankment pond's dikes are constructed above ground level. One disadvantage of this sort of pond is that it may require the use of a pump to fill it.
- **Excavated ponds:** A pond has been scooped out of the ground. This design has the disadvantage of requiring a pump to drain the pond.
- **Contour ponds:** The modest dikes of the pond are built with soil from the pond's excavation. The ideal location has a little slope (1-2 percent) so that the water supply channel can be built slightly above the pond water level and the discharge channel can be built slightly below it. There is no need for a pump because the ponds are filled and drained by gravity.

Barrage ponds: A dike is built over a natural watercourse to create a barrage pond. The ponds are thus mini conservation dams with the added benefit of being simple to build. However, controlling this system is challenging: it is difficult to keep wild fish out, and much of the feed given to the pond will be lost due to the stream. Only in exceptional circumstances does a properly constructed barrage pond overflow.

Pond Design

- **Size and Shape:** Although square and rectangular ponds are the easiest to construct, you can customize your pond to meet the size and shape of your property. A family pond of 300 m2 is a good size that can be built without the use of machinery. Although ponds can be much larger, it is preferable to have multiple little ponds rather than one large one for family use. You will also be able to harvest fish more frequently if you have multiple ponds.
- **Depth:** The shallow end of the pool is normally 30 cm deep, while the deep end is 1 meter deep. If the pond is used as a water reservoir during the dry season, it can be deeper. It is critical that the water be entirely drained in order to harvest it.
- **Types:** The type of pond you need to construct is determined on the terrain contours (topography). Flat and hilly locations require different sorts of ponds. Excavated ponds are constructed in flat regions by excavating an area as large as the pond requires. The water level will be lower than it was before. Contour ponds are constructed on a slope in hilly areas. The soil from the pond's top side is scooped up and utilized to construct a dam on the lower

side. Because the water level in the pond will be higher than the original ground level, the dam must be sturdy.

Steps for Building a Fish Pond

The most difficult and costly aspect of fish farming is constructing a pond. A well-constructed pond is a sound investment that will last for many years. The following are the steps of building a fish pond:

- **Prepare the site:** After removing trees, shrubs, and rocks from the area where the pond will be built, trim the grass in the area where the pond will be built. The length and width of the pond should be measured and staked off. Remove the top layer of dirt, which contains roots, leaves, and other debris, and dispose of it outside the pond. Save the dirt for later use on the pond dikes when grass is to be planted.
- **Build a clay core (in the case of contour ponds):** The basis for the pond dike is a clay core, which makes it sturdy and prevents water leaks. In contour ponds, a clay core is required and is constructed beneath those areas of the dike where the water will be above the original ground level. Excavated ponds do not require a clay core because the water level is below the original ground level.
- **Dig the pond and build the dikes:** Build the dike on top of the core trench with the soil you pulled out when digging the trench for the clay core. Avoid using sandy/rocky soil, as well as dirt that has roots, grass, sticks, or leaves. These will decompose over time, leaving a weak area in the dike where water might pour out. While constructing the dike, keep compacting the dirt at regular intervals. After each 30 cm of loose soil has been added, trample it thoroughly while spraying water on the dike. Then, using a hoe, a hefty log, or a piece of wood fastened to the end of a pole, pound it down. This will strengthen the dike.
- **Build the inlet and outlet:** A canal to bring water in, a silt collecting basin, and a pipe to transfer water into the pond make up the water intake. The water entering the pond often contains a lot of soil and silt, resulting in a murky pond. This soil will be kept out of the pond by a silt catchment basin. The soil will settle into this pit – termed a silt catchment basin – instead of entering the pond as a result of extending and deepening the inlet

canal just outside the pond dike. The water inflow line from the catchment basin flows into the pond via the pond dike. It should be around 15 cm above the water level to allow incoming water to splash down into the pond. This will keep fish from swimming into the input pipe and escape. It also aids in the incorporation of air (and consequently oxygen) into the water. Only in an emergency is the water overflow pipe used. The ponds should not have water flowing out of them daily. The overflow pipe removes surplus precipitation and run-off water from the pond during heavy storms. It is possible to install the overflow pipe at an angle. The screen will not clog with debris that may be floating on the pond surface if you place it with the intake underwater, as illustrated. Metal, plastic, bamboo, wood, or other materials can be used for the inlet, outlet, and overflow pipes. Install the pipes at the water's surface via the pond dike. Fish should not be able to enter or exit the pond through the pipes unless they are protected by screens. To keep wild fish and items like branches and leaves out of the pond, the INLET pipe is screened at the edge, which is outside the pond. To prevent fish from fleeing, the OUTLET (also known as drainage pipe) is screened inside the pond.

- **Protect the pond dikes:** Cover the pond dikes with the topsoil that was salvaged from excavating the pond when they're finished. Plant grass such as Rhode's grass (Chloris gavana) or star grass on the dikes (Cynodon dactylon). Plants with lengthy roots and trees should not be used since they will weaken the dikes and create leaks. The fertile topsoil will aid in the growth of fresh grass, which will also help to protect the dikes from erosion.
- **Fertilize the pond:** The application of fertilizers such as animal dung, compost, or chemical fertilizers can improve the natural fish food production in the pond. Before filling the pond with water, spread the fertilizer on the dry pond bottom. Fertilize the pond water on a regular basis, preferably once a day in the late morning or early afternoon. This continuous fertilizer application will maintain a steady supply of natural fish food. Before filling the pond, add lime or wood ashes to the pond bottom in addition to fertilizer if the soil is acidic. On water acidity, alkalinity, and hardness, use 10-20 kilogram of lime or 20-40 kg of wood ashes per 100 m2 of pond bottom.
- **Fence the pond:** A barrier around the pond will keep criminals

and predatory animals out while also protecting children from falling into the water. Plant a dense hedge around the pond's edge or construct a fence out of poles and thorn branches for a low-cost and durable fence.

- **Fill the pond with water:** Put rocks on the pond bottom where the water will land when it comes in from the entrance pipe before filling it. This will prevent the entering water from causing a hole in the pond's bottom and eroding it. Then fill the pond by opening the intake channel. Fill the pond gradually to avoid the dikes subsiding owing to uneven soaking. The water depth can be tested with a stick while the pond is filling. When the pond reaches the required depth, stop filling it. Do not overfill the pond to avoid it spilling. Because water moving through the pond slows fish growth by flushing away naturally produced fish food, the water in the pond should not flow through (and hence should be stagnant). Only enough water should be given to the pond to compensate for water loss due to evaporation and seepage. When new ponds are filled with water for the first time, they often seep because the soil absorbs some of the water. Continue to add more water for several weeks, and the pond should gradually begin to hold water.
- **Stock the fish:** Before stocking the fish, wait 4-7 days. This permits the pond's natural food production to reach a level that supports fish growth. If you opt to put substrates in the pond, you'll have to wait a little longer for them to be colonized by organisms that the fish can consume. As directed in, gently stock the tiny fish (known as fingerlings). It's important to note that the temperature of the water the fingerlings come from should be like the pond's water temperature. From here on out, it's critical to keep the pond in good shape and keep an eye on the water quality.

Periphyton-Based Fish Farming

Periphyton is a phylum of algae, bacteria, fungus, and other aquatic creatures that cling to hard substrates in the water. The slimy covering created by these organisms is referred described as a "periphyton mat." Fish production is higher in ponds with substrates, such as branches or bamboo poles put vertically across the pond, than in ponds without substrates, according to research. This method is called as 'periphyton-based fish farming,' and it was inspired by traditional brush park fisheries in natural waters, in which plants or branches are dispersed throughout the water body to attract fish and other animals.

Benefits of Periphyton-Based Fish Farming

- **Additional food:** One of the main benefits of putting substrates in ponds is that the submerged poles or branches quickly become colonized with a variety of microscopic organisms that the fish can consume. Food availability in the pond is raised in a natural way in periphyton-based fish farming, minimizing the need to fertilize the pond or give the fish with supplemental feed. This is critical, both economically and environmentally: extra feed and fertilizers can be costly, and the process is inefficient in any case, as most nutrients are lost as waste to the environment. When compared to fish from ponds where artificial feed or fertilizers are added, the fraction of nutrients preserved in harvested fish is significantly higher with periphyton (inorganic fertilizer, compost, manure, etc.). In periphyton-based ponds, fish utilize resources more efficiently. Because some species are more efficient at grazing on a three-dimensional structure like a bamboo pole (periphyton) than at filter feeding from the water column (phytoplankton = microscopic algae), this is the case.
- **Shelter:** Another advantage of adding substrates to the pond is that it protects fish from predators such as birds, frogs, and snakes. Although fish-eating perching birds can utilize poles to perch on, you can take precautions to prevent the birds from taking fish. Birds that pick fish from perches on poles, for example, are reliant on the height of the pole above the water column. Making the poles a little longer will make things more difficult for the bird. The abundance of sticks in the pond creates an impediment for diving birds, reducing the likelihood of predation. Human stealing can be reduced by placing poles or branches in the ponds, in addition to natural predators.
- **Fish health:** In general, fish survival is thought to be better in ponds with substrates than in ponds without substrates. Periphyton appears to have a beneficial influence on fish health, according to emerging data. It can function as an antibiotic against several disease-causing bacteria found in ponds, as well as a form of

vaccine for fish that eat it. Fish have also been seen rubbing against trees or poles to expel parasites.

Water Quality

The water quality in the pond is affected differently by different substrates. Bamboo, for example, is more resistant to dissolved oxygen depletion than quickly degradable organic substrates like sugarcane bagasse or paddy straw. Periphyton mats are also either oxygen producers (upper water layer) or consumers (lower water layer) depending on their position in the water column (bottom water layer). One can help to prevent oxygen shortages in the pond by managing the distribution of substrates in the water column. The periphyton mat captures suspended particulates, improving water transparency and, as a result, sunlight penetration into the pond. The periphyton mat also absorbs hazardous substances for the fish, such as ammonia and nitrate. Ammonia toxicity is a significant impediment to the expansion of pond-based fish farming. Bacteria that break down ammonia can colonize the surface of the substrates in periphyton-based ponds because of the well-oxygenated water column. These mats act as a biofilter, reducing ammonia levels.

Costs and Constraints of Periphyton-Based Fish Farming

In India, researchers calculated the expenses and revenues of a carp polyculture study in order to evaluate the economics of periphyton-based fish farming. Catla, rohu, and common carp were used in the trial. Sugarcane bagasse was employed as the substrate, with densities of 0, 7, 14, and 28 kg/100 m2. In all the experiments where substrates were utilized, fish production increased, but the increases were nearly identical in the 14 and 28 kg/100 m2 trials. As a result, the expenses associated with the 14 kg/100 m2 trial were utilized to compare to the costs connected with the trial without substrates. The additional costs for transportation, labor, and supplies for substrate installation came to Rs 5,960 (Indian rupees), but the increased revenue from fish sales came to Rs 24,500. The following are some of the major drawbacks of periphyton-based fish farming:

- Additional labor required for substrate installation and removal before harvest
- Possible conflicts in the use of the substrate in the household (as fuel or in other more productive activities)
- Cost of the substrate if this is not available on-farm

- Potential local deforestation if demand for substrates increases
- Problems with water quality if the system is not managed properly
- Insufficient knowledge of the biology of the system: fish species or species combination to be used, fish density, substrate type, density and so forth.

Maintenance and Monitoring

Regular maintenance and monitoring are required to attain a high level of fish production in the pond. The following tasks are performed daily:

- Checking the water quality (oxygen, pH, color, transparency, temperature, etc.)
- Checking the pond for possible water leaks
- Cleaning the screen of the water inlet and outlet
- Observing the fish while they feed: Do they eat normally? Are they active? If not, and if they are gasping for air at the surface, the oxygen level in the water is too low. Stop feeding and fertilizing and let water flow through the pond until the fish behave normally again. Otherwise, look for symptoms that could indicate a disease.
- Watching out for predators, or signs of predators such as footprints, and taking precautions if necessary
- Removing aquatic weeds growing in the pond

SNAIL FARMING

Introduction

Snail farming has probably been practiced for millennia, although no one knows when people started eating snails. Some sources suggest that snail farming has been done since the Roman Empire, although there is no evidence to back this up other than archaeological finds of mounds of snail shells dating back hundreds of years. However, because heli-culture is undoubtedly a Roman term, the connection with the Romans may explain why snail raising is named heli-culture. Snails, on the other hand, are a highly nutritious food that is high in protein, calcium, and iron, has no fat, and has a low cholesterol content. Vitamins, minerals, and two essential amino acids, lysine, and arginine, are found in their meat. Even though lysine is helpful to the body, we are unable to make it naturally, thus we must receive it through dietary sources. One of the best sources of this amino acid is snail meat. Arginine is another nutrient that aids in the formation of proteins in the body. It's found in red meat, fish, chicken, soybeans, grains, and dairy products, among other protein-rich foods. L-arginine can be found in snails. Most people who enjoy eating snail meat consider it to be a delicacy. Although some individuals avoid eating snails for religious or cultural reasons, it is safe to say that most people who try it adore it.

Snail farming has become a significant moneymaker and very profitable industry, especially because you can get started with very little money. It is one of the best ways to make money on a modest scale in a short period of time. Its technological, labor, and financial inputs are generally modest when compared to other types of livestock husbandry such as piggery, poultry, and cow raising. Many countries, particularly in Europe and Africa, now raise snails for both domestic consumption and exportation to other countries across the world, and the industry is rapidly expanding. Snail farming has the advantage of being able to be done on a small scale from the comfort of your own home. The farming method is simple and straightforward, and the space necessary for rearing can range from a few old tires stacked on top of each other, to utilizing enormous baskets or large clay pots, to building a little outdoor hut that is no more than 25 square feet (2.4 square meters). However, the size of your snail farm and whether it is for home use or commercial use will always be determined by the area you have available.

Benefits of Rearing Snails at Home

- A snailery, aside from crop cultivation, is completely silent. Snails are the only animal you can raise on a farm that is as silent as they are.
- Snail droppings have no odor. This is not the same as what you'd find in a piggery or a poultry farm. Snails can be kept near to or within dwelling quarters.
- The flesh is high in protein, phosphorus, calcium, and iron, as well as being flavorful.
- It's good for persons with heart problems because it's low in cholesterol, fat, and sodium, and it's a healthier alternative to red meat.
- The snail-rearing business has a large market potential.
- Cosmetics and other skin care products are also made with them.

- Reduced and eliminated cold sores, as well as the prevention of lysine deficient symptoms, are some of the potential health benefits of lysine (found in snails).
- Snail flesh contains arginine, which aids in immunological and hormone function, dilates, and relaxes the arteries, supports a healthy renal system, and aids wound healing.

Edible Land Snail Species

All land snails are edible if cooked properly. Their flavor varies by species and cooking method, and tastes may differ by culture. Although all lands nails are edible, only a small percentage are deemed tasty, and even fewer are economically valuable. Only a few species with economic worth are suited for productive farming. The edible lands nails that are well-known for their culinary value are discussed below. Edible land snails' range in size from a few millimeters to huge African snails that can reach 312 mm (12+14) in length on rare occasions. Although other types of snails are consumed, the term "escargot" usually refers to Cornu aspersum or Helix pomatia. The huge African snail Lissachatina fulica is sliced and canned and sold as escargot to select consumers. Although terms like "garden snail" and "common brown garden snail" are somewhat useless because they relate to so many different sorts of snails, they do occasionally refer to H. aspersa.

- **Cornu aspersum:** The Helix aspersa Müller, formerly known as the Helix aspersa Müller, is also known in French as the «petit gris," "little grey snail," "escargot chagrine," or "la zigrinata." A mature adult's shell has four to five whorls and is 30 to 45 mm in diameter. It is found along the Mediterranean coast and along the coasts of Spain and France. It can be found on several British islands, where it was introduced by the Romans in the first century AD (Some references say it dates to the Early Bronze Age.) It was transported to California by the French in the early 1800s, where it has become a severe problem. These snails are now found across the United States. Before 1850, it had been imported into several Eastern and Gulf states, and it was afterwards spread to other nations such as Australia, South Africa, New Zealand, Mexico, and Argentina. The lifespan of C. aspersum is 2 to 5 years. This snail is found in woodlands, fields, sand dunes, and gardens, and is more adaptable to varied seasons and situations than many snails. C. aspersum's adaptability not only expands its range, but it also makes growing it easier and less risky.

- **Helix pomatia:** The shell measures roughly 45 mm wide. The "Roman snail," "apple snail," "lunar," "la vignaiola," the German "Weinbergschnecke," the French "escargot de Bourgogne" or "Burgundy snail," or "gros blanc" are some of its other names. It lives in wooded mountains and valleys up to 2,000 meters (6,600 feet) altitude, as well as vineyards and gardens, and is native to a significant area of Europe. It's possible that the Romans brought it to Britain. It was brought to the United States by immigrants in Michigan and Wisconsin. Many people consider H. pomatia to be the "escargot par excellence" because of its flavor and bigger size compared to H. aspersa. The H. pomatia, on the other hand, has not proven to be economically feasible for farming.

Figure 21: Roman Snail (Helix pomatia).

Source: https://crateandbasket.com/types-of-snails/

- **Otala lacteal:** The "vineyard snail," "milk snail," or "Spanish snail" are all names for the same mollusk. The shell measures 26 to 35 mm in diameter and is white with reddish brown spiral stripes.

Figure 22: vineyard snail (Otala lacteal).

Source: Google Images

- **Iberus alonensis:** The shell of the Spanish "vaqueta" or "serrana" measures about 30 mm across.
- **Cepaea nemoralis:** The shell of the "grove snail," or "vaqueta" in Spanish, measures around 25 mm across. It is native to Central Europe and has since become naturalized in several US states, ranging from Massachusetts to California and from Tennessee to Canada. Its habitat is diverse, ranging from woodlands to dunes. It eats mostly dead plant material, but nettles and buttercups are favorites, and it will also eat dead worms and snails.
- **Cepaea hortensis:** The "white-lipped snail" has a shell that is about 20 mm across and often has pronounced dark stripes. Its natural habitat is across central and northern Europe. In colonial times, this plant was introduced to Maine, Massachusetts, and New Hampshire, but it never became established. The habitat of C. hortensis varies, although it prefers colder and wetter climates than C. nemoralis. C. hortensis and C. nemoralis are less popular than larger European land snails due to their smaller size and some people's perception that they don't taste as delicious.

Figure 23: White-lipped snail (Cepaea hortensis).

Source: Google Images

- **Otala punctata:** The shell measures about 35 mm across and is known as "vaqueta" in some parts of Spain.
- **Eobania vermiculata:** The "vinyala," "mongeta," or "xona" is around 25 mm in diameter. It was imported into Louisiana and Texas from Mediterranean countries.
- **Helix lucorum:** The Turkish snail, so named because of its widespread distribution in Turkey, is around 45 mm across the shell. It can be found in central Italy, as well as in Yugoslavia, Crimea, Turkey, and the Black Sea region.
- **Helix adanensis:** comes from around Turkey.
- **Helix aperta:** It's around 25 mm long. Its meat is greatly sought after. It originated in France, Italy, and other Mediterranean countries, but it has now spread to California and Louisiana. It's also known as the "burrowing snail" since it only comes out of the ground when it's raining. It burrows three to six inches into the earth in hot, dry weather and goes dormant until rain softens the soil.

- **Theba pisana:** The "banded snail," also known as the "cargol avellanenc," is around 20 mm long and dwells in dry, exposed areas, usually near the sea. It originated in Sicily and has since spread throughout Europe, including England. This snail is a major horticultural pest, and it is the same "white snail" that was previously exterminated in California by using flamethrowers to destroy entire areas. It can ruin a garden in 24 hours and a citrus or other crop in a couple of nights if it has a big population of up to 3,000 snails per tree.
- **Sphincterochila candidissima:** or Leucochroa candidisima, the "cargol mongeta," or "cargol jueu" measures about 20 mm.
- **Lissachatina fulica:** Giant African snails, Lissachatina fulica and other members of the group Achatinidae. They can reach a length of 326 mm (1 ft 34 in). Their original range is in East Africa, south of the Sahara. In 1847, this snail was purposefully introduced into India. In 1925, there was a failed attempt to develop it in Japan. It was accidently discharged in California after WWII, Hawaii, and North Miami, Florida in the 1970s. It is a devastating agricultural pest in many areas, causing significant crop loss. Its slime and feces, as well as the stink that arises when something like poison bait kills a huge number of people, are also a nuisance due to its size. The United States has made significant efforts to eradicate Achatina (previously Achatina fulica; now Lissachatina fulica). The importing and possession of live Giant African snails is prohibited by the United States Department of Agriculture (USDA). However, because of the bright "tiger stripes" on their shells, they are still sought after as pets. Giant African snails can be farmed, although their requirements and farming methods are vastly different from those used for Helix species.

Figure 24: Giant African Land Snail (Achatina fulica)

Source: https://crateandbasket.com/types-of-snails/

Techniques for Snail Rearing

When you eventually decide to establish a home-based snail farm, you'll discover that depending on the type of housing you have available, your costs can be surprisingly minimal. Set up a snail house in a corner of your garden if you have one. If you don't have a lot of room, the tire-stacking option is the best choice. If you reside above ground level, such as in a seventh-floor flat, you have the option of living in a clay pot or a basket. Begin with the procedures outlined below after you've identified the ideal rearing structure for you.

Set Up an Outdoor Snailery

The most crucial aspect of a snail's environment is the soil. The soil must be moist and rich in organic matter to promote their growth and development, like that which supports the cultivation of leafy greens, tomatoes, and carrots.

- Clear the ground of weeds, bushes, and twigs, and make sure the texture is loose enough for mature snails to dig readily into the soil to deposit their eggs and hibernate during the dry season.
- Hand-till (plough) the soil to loosen the grains and provide a snail-friendly texture.
- Make a vegetable and leguminous plant paddock.

- Surround the paddock with a fine-mesh or netting enclosure. Keep in mind that snails can slip away unobserved.
- In the paddock, plant leafy greens, legumes, coco-yams, bananas, and dwarf pawpaw plants and wait for them to flourish.
- Introduce your breeders to the enclosure and make sure the paddock is well-watered. It will offer the snails with both shelter and nourishment.
- Pests such as crawling insects and other predators pose the greatest threat to snails, particularly in tropical areas where they are commonly reared. Build a gutter around the housing and fill it with water and pesticide to keep them out.

Creating an Indoor Snailery

Snail rearing in an interior environment, such as a patio or veranda, necessitates the use of objects such as old tires, drums, hutch boxes, clay pots, and baskets, as previously described. They must have a soft enough humus soil or compost for the breeders to burrow into and lay and bury their eggs. You can harvest the eggs once they've been laid and set them in smaller containers filled with damp, loamy soil. The hatchlings will be able to grow unhindered with the help of these incubation boxes.

- System of stacked tires: Discarded tires can be used as snail cages at a low cost. Place four old tires in a stack and sandwich chicken wire mesh between the top and bottom. Fill the bottom two tires with loamy or humus soil, and keep the enclosure to a maximum of five breeding snails.
- Drums that are empty Snailery: Make drainage holes at the bottom of oil drum pens before filling with appropriate soil to a depth of around 10 inches (25 centimeters). Cover the drum with a framed and removable wire mesh lid. In each container, place around five or six breeding snails. At all times, the soil must be kept moist.
- Hutch boxes, single, double, or triple, are boxes on stilts with drain holes in the bottom, filled with 20–25 centimeters of soil, and put waist-high for easy handling and moving of eggs or hatchlings from one pen to the next. To keep insects and other pests from crawling up the crate legs and attacking the snails in the hutch, aprons made of plastic (or old cans) must be attached to the legs. Nylon mesh and wire netting should be installed in the lids, which are a wooden frame hinged to the container.

- Baskets and pots made of clay: This is the simplest and most cost-effective approach to start a snailery. If you're going to use this strategy, you're probably only going to raise snails for your own food. The main disadvantage of this procedure is that you must regularly clean out the polluted soil to avoid foul odors and infected snails. Fill the container with sieved black soil (no more than 25 centimeters (10 inches) and a removable framed wire mesh lid, just like the other methods. Make sure you only add three breeders.

The housing system for your snail farm, regardless of its size or type, must match the following requirements. It must be:

- Snails are master escapists, and if you don't stop them, they'll swiftly travel all over your (or your neighbors') garden or house.
- roomy, in line with the snails' developmental stage (hatchlings, juveniles, breeding snails, or mature snails fattened for consumption). Overcrowding is a problem for snails, as it impedes their development and raises the risk of disease. Hatchling rearing densities range from 100 per m2 to 7-10 per m2 for breeding snails.
- For managing the snails, placing feed, cleaning, and other duties, it is easily accessible and easy to operate in or with.
- Insects, predators, and poachers are all well-protected. Snailearies can be made from a variety of materials, depending on cost and availability.

Snails Feeding

Protein is required for growth and development, carbohydrates are required for energy, calcium is required for healthy shells, and other minerals and vitamins are required. As a result, whatever you feed them must have these nutrients. The good news is that snails are herbivores and can eat a wide variety of plant meals, except for harmful chemicals-producing plants and waxy or hairy leaf flora. Young snails should be fed fragile leaves and shoots, while older, mature snails should be fed leaf litter and rotting fruits. The following diets and nutritional supplements are recommended:

- Plant leaves such as cocoyam, papaya, okra, eggplant, loofa, cabbage, lettuce, and cassava
- Papayas, bananas, strawberries, pears, mangos, oil palm, figs,

eggplants, watermelons, tomatoes, and cucumbers are just a few examples of fruits.

- Sweet potatoes, coco-yams, yams, carrots, and other types of potatoes are tubers.
- Bananas, pawpaws, yams, plantains, carrots, and pineapple peels are common household peels.
- Cooked rice, beans, and peas that haven't been salted.
- Snail feed should also contain powdered calcium (for optimum shell formation) derived from eggshells, wood ash, ground limestone, crushed oyster shells, or bone meal.
- Vitamins D, E, and K are present in modest levels in supplemental vitamins and minerals. Sunflower, copra cake, wheat germ, lettuce, cabbage, and spinach are all good sources. Mineral sources can be added to the pen by putting licking stones holding the necessary minerals.

Feeding should be done twice a day, in the morning between 7 and 9 a.m. and in the evening between 5 and 7 p.m.

Snail Breeding

Snails in captivity suffer from overcrowding, which has a negative impact on their growth and development. They will grow slowly, will be underweight, will lay fewer eggs, and may not breed at all.

Starting Small: If you're new to snail farming, it's better to start modest, with only a few breeders. You'll learn their behaviors and get better at snail rearing as you become more comfortable with your home-based heli-culture. Then, if needed, you can raise their numbers. 1–1.5 kilograms per square meter, or 15–25 snails per square meter, is the acceptable density.

Gestation Process: When you receive or buy breeders, they are almost certainly all full of viable eggs. Between fertilization and egg laying, there is a one-to-two-week gestation period. They burrow into the earth in the breeding pen and deposit eggs in clusters, with an average of eight eggs per cluster. Any eggs discovered on top of the loam must be buried as soon as possible. Mature snails must be taken from one pen and transported to another after the eggs have been laid, leaving the eggs to gestate. The incubation period varies depending on the environment, but it normally takes 25–35 days.

Protecting Hatchlings: The eggs hatch and slither out of the ground. Remove them from the 'birthing' pen and place them in a nursery pen. A smaller clay pot, basket, or two stacked old tires can be used for indoor snail raising. Because newborn snails are sensitive to dryness, the environment must always be maintained moist and clean.

Snails Harvesting

The age and size at which snails should be ready for harvesting in a small-scale company like snail farming is determined by the farm's primary goal: whether it is to rear food for personal use or to sell. You can pick when to harvest snails if you're breeding them for personal use. The timing will depend on whether you prefer small, tender snails or huge, meaty ones. Buyer preferences decide the age and size of harvesting for snail breeding for sale. These preferences differ depending on where you are in the world. Snails, on the other hand, require an average of 12 months to attain a size and weight that is suitable for feeding. Snails can be harvested when they are between the ages of 12 and 18 months because their growth rate slows after that. Because they crawl out of the nooks and crevices of the pen at night, harvesting is best done at night. The greatest time to find and select them is at night, when they are most active.

The Lifecycles of Snails

Snail farming for consumption can be done at home with just two snail pens. You'll need at least three snail housings if you want to sell them, but this only works for outside snail houses. Hatchlings, growers, and breeding snails are kept in different crates or pens by semi-intensive farmers. With that in mind, let's look at the four major stages of a snail's lifespan and how best to care for them.

- **Eggs:** Domesticated snails have the advantage of continuing to lay eggs even during dry seasons if given the correct conditions. They also deposit eggs in clusters of eight (about) and bury them in the ground. Any eggs discovered on top of the loam must be buried as soon as possible. The time it takes for the virus to incubate varies, but it normally takes between 25 and 35 days.
- **Hatchlings:** These snails demand a higher level of humidity than adult snails. Water should be provided at regular intervals to keep the dirt in their pens moist. The density of snail hatchlings and juveniles is roughly 100 snails per square meter.

- **Growers:** Growers should be separated into separate pens when they are around three months old, with a stocking density of 30–40 snails per square meter of soil surface.
- **Breeders:** When they reach sexual maturity, which occurs around the age of 10–12 months, they begin to deposit eggs. Breeders must be moved to boxes or enclosures containing no more than 15 snails per square meter. They can be kept in fattening pens until they are suitable for consumption or sale after they are no longer needed for breeding. Breeder snails are allowed to develop until they are 18 months old.

Purged, cleaned, packaged, and marketed in a refrigerated sleeping state are medium-size snails (used for European-style cooking). Snails are mainly sold live in open markets in West Africa. West Africans feel that cooking frozen or refrigerated snails destroys their natural flavor. It's all a matter of personal preference. Snails for home consumption require six months to grow out on average, while those sold at markets take eight months to reach marketable age.

Tips for Snail-Rearing Farms

- If you're farming on a balcony indoors, make sure the snail housing has a good drainage system, or make a basic drainage system if you're rearing the snails outside.
- The snail house should be protected from the sun, rain, and strong winds.
- Clay and sandy dirt should never be used in the housing since they are harmful to the incubating eggs. Loamy soil or humus is the greatest form of soil to use. Snails with weak shells and slow growth will be the result of a lack of access to healthy soil.
- Changes in the soil composition will occur when the pen's soil becomes polluted by mucous and droppings. Change and replenish the soil on a quarterly basis to avoid this.
- Snail farming should ideally begin around the start of a wet season, as this is when snails begin to reproduce.
- Snails of medium size should be used for breeding. Snails of this size are too old to reproduce.

- The snails you choose as breeders must be sexually mature and have healthy shells. They should be between 100 and 125 grams (0.2 and 0.27 pounds) in weight.
- Overcrowding the pen or containers is not a good idea. Overcrowding has an impact on their growth and development, as well as encouraging cannibalism. Allow five snails per square foot (or 50 snails per square metre) of soil surface.
- Snails require a variety of foods because they prefer to eat what they like—a mix of fruit peels, leafy greens, and tubers should provide a nice selection for them.
- Snail shells can be painted and used as tabletop ornaments, planters for small decorative plants (like cacti), or as an inlay for lamp bases, ceramic pots, or to embellish bottles and containers if you are creative or into DIY activities.

Chapter 4

Small-Scale Farming Profit Making Options

CONTENTS

Overview....180
Small-Scale Machinery....181
Soil Health for Small-Scale Farmers....185
Permaculture....190
Forest Gardening....195
Aquaponics....199
Smart Water Management....215
Integrated Pest Management....224
Diversify the Smallholder Farm....227

OVERVIEW

Farming profit margins are increased using tactics like those used in other businesses around the country. Farmers strive to maximize work, cut costs, and increase revenue in order to keep as much profit as possible. There are a lot of ways farmers can use to enhance profit margins without affecting the quality of the products they bring to market, other from just raising pricing. Farmers can reduce product shipping expenses and enhance profit margins by selling to local community markets and grocery shops. A farmer selling produce or meat to local grocery stores and markets doesn't have to travel as far to reach delivery destinations, saving money on gas and reducing wear and tear on delivery equipment. In addition, a farmer can save money on payroll by having shorter delivery routes for his drivers. Because an increasing number of people are eager to lessen their carbon impact by purchasing locally grown products, local grocery stores prize produce and meats from local farms.

A farmer can cut energy expenses and boost profit by investing in alternative energy alternatives, such as wind power. According to the Union of Concerned Scientists, a nonprofit advocacy group, wind turbines can provide a cheaper form of energy to power farm equipment than traditionally generated electricity. A farmer can also lease property to wind companies, who will erect turbines on the location in exchange for monthly fees. According to the Union of Concerned Scientists, land lease payments can add $2,000 to $5,000 to a farmer's annual income for each installed wind turbine. Crop rotation tactics can help farmers save money on fertilizer, enrich their soil, and increase their take-home pay. Planting soy beans while corn isn't in season, for example, boosts nitrogen levels in the soil. According to the Environmental Protection Agency, these measures can also help kill hazardous insects and slow the growth of harmful microorganisms. According to the EPA, Integrated Pest Management can assist a small or large-scale farmer in bringing a higher percentage of produce and other food products to market.

Because of the higher yield and healthier soil, a farmer can keep a higher percentage of his earnings as profit. Investing in more efficient farm equipment, such as tractors, harvesters, and grain separators, can lower a farmer's long-term equipment expenditures and improve earnings. A farmer may incur higher upfront costs to purchase more energy-efficient equipment, but the benefits in decreased fuel and maintenance costs over time may outweigh those initial costs. Investing in new capital assets that are intended to provide income for the farm helps a farmer to repay these costs over time through depreciation. This ensures that the farmer is entitled for tax deductions for the duration of the items' useful lives.

SMALL-SCALE MACHINERY

Introduction

By increasing production efficiency, the use of appropriate agricultural equipment and instruments for small-scale intensive crop production contributes to the farm's viability. Plant propagation, soil preparation, planting, pest and weed control, irrigation, harvesting, postharvest processing, storage, and distribution all require tools and equipment. Sustainable agriculture is a labor-intensive business, and farmers can boost earnings by increasing crop yields, improving crop quality, and lowering expenses by choosing the right tool for the job. The location and growth circumstances of the farm, the type of crops being grown, the production procedures being employed, and how the crops will be marketed are all factors to consider when selecting proper agricultural equipment and tools. Previously, the size of the business—or the farm—was the most significant determinant in making a profit. Many small-scale farmers today make a lot of money on land that is only five acres or less. Practical farmers of this scale can manage their production and farm finances in a sustainable manner, thanks to modest annual operating expenses and appropriate capital needs. Farms that produce intensely on fewer than five acres rely on the variety of their manual labor, but they may also use mechanical equipment to increase output efficiency. As farming systems migrate from hand labor to tractor labor, tools and equipment should correspond to the volume of production. Compromises are required as farming systems transition from hand labor to tractor labor. When scaling up production to incorporate a tractor and equipment, reducing the number of possible row spacings, limiting the number of different bed widths, and constructing beds that include slope, soil erosion, and tractor-wheel widths are all key concerns. It's also crucial to think about how to best match the correct equipment to the cropping system, considering things like how the soil and plants are handled, as well as how cover crops and crop rotations are used. Diversified cropping systems can make things more complicated because different crops require varying row spacing, therefore equipment may need to be changed for each crop. Small-scale intensive crop production equipment is typically simpler and less specialized than larger-scale production equipment. As a result, the equipment is frequently less expensive and requires less investment. It may appear that the economics of owning and maintaining farm machinery are complex and even daunting. For some farmers, renting or borrowing equipment may be a possibility; however, investing in adequate and practical equipment that matches the whole-farm cropping strategy can result in a modest cost per hour increase. Even after costs like driver labor, gasoline, maintenance, and equipment depreciation are accounted in, the investment pays off in labor savings. Market vegetable farms adopting sustainable farming approaches invest between $2,011 and $26,784 per acre in equipment, according to a 2005 report released by the Center for Integrated Agriculture. Due to the high expense of new equipment and the fact that new equipment available in the United States is no longer built for small-scale production, small-scale farmers prefer to invest in old equipment. Many small-scale farmers are sourcing new equipment from international manufacturers to satisfy their production needs, working with local distributors, and incurring the price of having it imported or constructed locally. Making a profit after the original expenditure of purchasing the equipment is dependent on the accompanying costs of maintaining, storing, and repairing it, regardless of where it is created. Although draft-animal power has a place on many sustainable small farms, there is no mention of draft-animal power equipment in this paper. Many of the devices mentioned in this publication, however, were built for horses in the first place and may be utilized with both draft and tractor power. For more information on the usage of draft animals, see ATTRA's Draft Animal Power for Farming.

Mechanization for Smallholder Production

Smaller farms, on the other hand, tend to misuse labor, resulting in labor-intensive farming technology and low labor productivity. Furthermore, with the changing demographics in agriculture, labor is becoming increasingly scarce. Smallholder farmers' lack of access to more modern types of farm power is a crucial element that perpetuates labor hardship, particularly among women (e.g., weeding, threshing, shelling, and head-loading), while limiting production and farm output. The lack of proper machinery that caters to and matches the needs of small-scale farms is one of the key causes of low worker productivity. The lack of adequate technologies has resulted in a low rate of adoption of mechanization among smallholders. The relevance of adapting mechanization technology to farm features is highlighted by the problem of enhancing labor productivity among smallholder farmers. The adoption of four-wheel tractors (4WT) is not economically viable due to the prevalence of small-scale farms and their limited land sizes. The scarcity of area for fodder production owing to declining land sizes, as well as the consequences of drought and increased animal illness risk, are threatening the choice of animal traction. Alternative and proper mechanization methods are critical in this situation. The adoption, use, and diffusion of smaller scale equipment such as two-wheel tractors (2WT), shallow tube-wells, smaller scale low lift pumps, and 2-, 3-, and 4-wheeled motor transportation vehicles resulted in the Green Revolution in Asia. The versatility and capacity to execute many duties, such as planting, ripping, hauling, shelling, threshing, and water lifting, make the 2WT an excellent choice for mechanizing the various functions along the agricultural value chain. However, because it is extremely difficult to separate mechanization for profitable on-farm operations from other rural economic tasks, the usage of two-wheel tractors should be viewed as a rural mechanization strategy rather than an agricultural mechanization approach. The 2WT technologies have the advantages of being more suitable for smallholding sizes, being able to traverse small fields and plots more efficiently, and being multi-functional. Furthermore, they are better suited to reducing on-farm drudgery, especially for women. Two-wheel tractors are also energy efficient, which is important for promoting agricultural intensification in a sustainable manner and increasing output with minimal environmental impact. The need for higher efficiency in the use of resources for farming and greater energy savings are prevalent themes in this strategy. This can be accomplished most effectively by encouraging two-wheel tractors, which improve farm power balance by increasing power supply, while also supporting sustainable intensification principles, which reduce power demand while increasing power supply through appropriate mechanization. Only by appropriately addressing the issue of dwindling agricultural power can sustainable intensification be achieved. Improving access to energy-saving machinery is referred to as "smart mechanization." As previously stated, one of the driving forces behind the acceptance and use of two-wheel tractors is the draught power issues. A switch from animal draught to tractor power has the added benefit of decreasing strain on crop leftovers for animal fodder and increasing the amount of surface mulching available. Conservation is being practiced. Agriculture methods are especially significant for reducing electricity consumption and mitigating soil degradation by avoiding soil inversion through low and zero tillage practices. 2WT-based CA is most likely to improve food security by lowering cultivation costs, boosting yields (especially in upland areas with small grains), and, as previously stated, enhancing labor productivity. The following are some of the tasks done by two-wheel tractors:

Figure 25: 2 Wheel Tractor

Source: Google Images

Tillage: In terms of time and labor, employing two-wheel tractors for land preparation, planting, and tillage has significant economic advantages over using a manual hoe or draught cattle power. On sandy to loamy soils, a 2-wheel tractor can tear one hectare every day while using about 15 liters of diesel.

Shelling/threshing: Shelling and threshing is a time-consuming procedure that requires a lot of labor in post-harvest crop management. Human labor is typically used for shelling/threshing, especially in maize, beans, and pigeon 3 Mechanization for smallholder production Shelling and threshing have been more efficient and cost-effective since the introduction of two-wheel tractors.

Pumping/irrigation: Water lifting and irrigation are also possible with two-wheel tractors, particularly in rice and vegetable farming. Because of the availability of furrow irrigation systems, the use of irrigation in maize production is expanding in various East and Southern African countries. However, farmers require pumps and power to support the irrigation system, and the 2WT is adaptable in covering this operation.

Milling: Two-wheel tractors can also mill rice and grain, and in several SSA nations where two-wheel tractors are more common, their use in milling is increasing due to high demand among farmers. The usage of two-wheel tractors improves efficiency because fuel consumption is low.

Transportation: Tractors with two wheels are also widely utilized as modes of transportation for both productive and recreational purposes. Tractors enable farmers to carry agricultural and non-farm produce to local markets, as well as purchased inputs from dealers to the farm. In most nations, ox carts, motorcycles, and bicycles are used to transport inputs and retrieve water, but tractor transport is in high demand. It is critical that any technology adopted is appropriate for meeting the genuine needs of farmers, that it can be used efficiently and effectively, and that it is economically viable. Increasing mechanization does not necessitate big expenditures in tractors and accessories, but rather a transition to a different combination of land, capital, and labor that results in higher farm revenue by increasing productivity, lowering costs, or a combination of the two.

SOIL HEALTH FOR SMALL-SCALE FARMERS

Introduction

Healthy soil provides food for plants, stores nutrients for them, and provides important ecosystem services including water purification and carbon sequestration. According to studies, soil is the most important storehouse of biodiversity and the most valuable single natural resource, valued at about $4 trillion in 2012. Soil is also important in preparing for and preventing climate change. The United Nations Food and Agriculture Organization designated 2015 as the International Year of Soils to highlight the vital role soil plays in our lives. In contrast, bad soil can have disastrous consequences for everything from food security to global commerce to environmental quality. Smallholder farmers in Sub-Saharan Africa and South Asia, where the land is immensely diverse, chronically devoid of key nutrients, and sustains some of the world's highest population densities, are particularly vulnerable to nutrient-deficient soils. Smallholder farmers—those who cultivate fewer than several hectares of family land in tropical and subtropical developing countries—are hampered not just by poor soils, but also by a lack of access to high-quality seed and fertilizer, which the Western world has had for over a century. As a result, farmers are the largest group of individuals living in abject poverty in these areas. Governments are aware of the irony of hungry farmers. Over the last two decades, for example, development priorities in Sub-Saharan African countries have centered on agriculture to bring about the kind of Green Revolution that bypassed this continent the first time around and contributed to rapid economic growth in places like Mexico and India. The technologies that are required to generate healthy, resilient soil systems have been established for decades or millennia. Getting proven soil health treatments to the last mile at scale has proven more difficult. Despite decades of research establishing the relevance of soil health in smallholder farming and the usefulness of soil management practices, there are currently few regional examples of successful large-scale soil health programs. Conservation agriculture (CA), for example, has been widely pushed as a long-term soil management strategy for decades. Reduced tillage, crop residues as soil cover, and crop rotation are the three pillars of CA. While many of the CA components have been implemented by farmers in North and South America, and the benefits have been well established in scientific literature, adoption rates in smallholder farming systems have been extremely low. Another example is fertilizer deep placement (FDP). FDP, which was first developed in the 1980s for Asian rice systems, entails inserting compressed fertilizer briquettes deeper in the soil (>10 centimeters) near the roots of transplanted rice. When done correctly, FDP can dramatically enhance rice grain yields while lowering fertilizer requirements. Despite its well-documented advantages, FDP is not commonly used.

Ways To Improve Soil Health

Increase the amount of plant and animal matter going back onto fields: Organic matter in the soil is an important component of a healthy soil. The key to soil health is obtaining high levels. On many farms, levels are low or dropping; we must work together swiftly to reverse this trend by ensuring that farms recycle more plant and animal matter back into soils.

Improve soil health monitoring: Analyzing soils is an important first step in making informed decisions about soil health, but some farmers fail to do so on a regular basis.

Encourage soil organisms: both those that add to the soil's structure and those that release nutrients the current emphasis on chemical inputs began when we didn't fully comprehend the significance of soil life. We are only now discovering the importance of soil biology in agriculture. To improve our ability to cope with floods and droughts and increase crop yield, we must discover how to best nurture this life.

Encourage soil organisms by doing the following:

- Compost is a great addition to any garden. Microorganisms require a lot of organic materials to flourish since carbon is their primary energy source. Compost helps ensure that your garden has a consistent supply of organic materials. Check out our composting advice.
- Cover crops should be planted. Living soil need food, which is provided by cover crops. Learn how to choose the proper cover crops for your garden by reading our cover crop tips.
- Ensure that your soil is well-watered. Microbes grow in moist, moderately temperate soils. High temperatures can destroy microorganisms if the soil is left exposed and dry.
- Physical disruptions should be avoided. When the soil is disturbed, such as by tilling or compaction, the structure of the soil and the habits of the microorganisms living inside it are disrupted.
- Mulch your flower beds. Mulching with organic matter such as leaves or pine needles helps the soil retain moisture while also adding organic matter.
- Pesticides should be avoided. Chemical pesticides, herbicides, fungicides, and fertilizers have the potential to wipe out entire populations of soil microorganisms. Instead, use organic fertilizers like compost and manure, which slowly release nutrients into the soil and do not leave residues like many chemical fertilizers.

Cover up bare soil with continuous plant cover: Healthy soil is hidden by plants, so you can't see it. Plant roots bind soils together, preventing erosion and allowing air to pass through the crevices between roots. Through plant-fungal interactions, roots also promote better soil communities. However, the benefits extend far beyond the farm, with significant gains in biodiversity, carbon storage, flood and drought control, and water quality. The following are some of the options:

- Under Agri environment options/ Ecological Focus Areas, convert susceptible land to permanent grassland.
- Cover crops, green manures, and under-sown crops should all be used whenever possible to improve soil fertility.

Bring more trees onto farmland: This can be fruit trees. Available practices are:

- Increase your understanding on the value of trees

Reduce soil compaction from machinery and livestock: Soil compaction is a major problem it can resulting in increased surface run-off, drought stress, fewer grazing days, poor root growth, and lower overall yields The following are some of the options:

- Conduct a routine visual inspection to detect levels of compaction on the farm and take corrective action. To lessen the risk of compaction, a variety of mitigation techniques are available:
  - Use lighter machinery
  - Adopt Controlled Traffic Farming (arable farmers)
  - Use correct tyre pressure
  - Reduce the number of passes
  - Try out reduced-till/ no till farming
  - Avoid over-grazing
  - Avoid trafficking over and grazing wet land

Design crop rotations to improve soil health: Farmers are under pressure to make short-term decisions due to the present tendency in farming for simple arable crop rotations, financial difficulties, and the growing trend for short-term rentals. As a result, the UK's food diversity is dwindling, weeds are out of control, and soils are being depleted of nutrients and organic matter. To counteract this trend, we must ensure that farmers can create the diverse, long-term crop rotations that our soils require. The following are some of the options:

- Many high-value crops, such as potatoes and carrots, cannot be cultivated sustainably in a short cycle. Create a longer crop rotation to avoid these issues. Instead, then focusing on annual margins, consider financial margins throughout the length of a cycle.
- Before returning to the same crop, carefully plan your crop rotations with more crop variety and longer breaks. More focus should be placed on crops like legume catch crops that assist to protect soils and build soil organic matter.
- To take advantage of the soil's variable nutrient profile, plant crops with diverse rooting depths.
- Include temporary leys, which are used for grazing by cattle, in your rotation whenever practical.

Farming Systems that can Improve Profits

You must know where to begin if you want to raise your farming output and earnings. The following are several systems that, if properly implemented, can boost the profitability of a small-scale farm.

Double Cropping

Multi-cropping is a collection of techniques that includes double cropping. Multi-cropping refers to the various ways farmers might use one piece of land over the course of a single season or calendar year. Double cropping, cover cropping, integrated crop-livestock systems, and woodland-based systems including woodland pasture and agroforestry are the four main types of multi-cropping. While they all have financial and environmental consequences, they differ significantly. Double cropping is the harvesting of two crops or commodities in the same calendar year, such as winter wheat in the spring and soybeans in the fall, according to a strict definition. Some vegetable crops are double cropped if sown many times per year, however crops that are sown once and harvested multiple times (such as hay) are usually classified as a single crop with yields recorded to account for the repeated harvests. Planting two crops but only harvesting one is known as "cover cropping." Cover crops are commonly employed to preserve soil moisture and integrate organic matter back into the soil, among other environmental benefits. Harvesting one crop and then putting livestock to feed on residue or a second crop are referred to as "integrated crop-livestock systems." Livestock graze "dual-purpose crops"—a typical form of crop-livestock system—either before or after crop maturity. Winter wheat and winter rye are examples of dual-purpose crops. Any method that involves trees or tree products is referred to as "woodland-based multi-cropping." Agroforestry is a unique land management technique that combines agriculture and forestry with the goal of improving productivity, profitability, and environmental stewardship. A popular land use is "woods pasture," which uses woodland for grazing and pasturing livestock while timber products grow.

Economic Benefits and Barriers to Double Cropping

The growing interest in double cropping stems from both economic and environmental considerations. Increases in double cropping may provide other environmental or economic benefits in addition to avoiding the negative environmental repercussions commonly connected with agricultural development. Double cropping, for example, might minimize fertilizer requirements in particular situations. Double cropping can also help prevent soil from wind and water erosion by decreasing soil exposure between harvest times. In other circumstances, however, double cropping may need the use of additional inputs such as pesticides, herbicides, fertilizer, irrigation water, or conventional tillage, all of which have severe environmental consequences. The potential for increased revenue from a second crop on the same land base is one of the economic benefits of double cropping. Selling a second commodity could result in higher profits, especially if fixed costs (such as tractors) can be spread over greater production (Beuerlein, 2001). Furthermore, technological advancements such as early maturing seed varieties and no-till equipment may make double cropping more economically feasible. The relatively low acceptance rate for double cropping, on the other hand, shows that economic hurdles are limiting its use. Although a second crop might increase revenue, it frequently reduces per-crop yields due to the possible reduced growing time available for each crop, lowering the farm's overall annual revenue potential. Furthermore, adding a second crop may increase production expenses, and in some cases, double cropping may be riskier than raising a single crop. Double cropping, for example, may be more subject to moisture and weather fluctuations during the growing season, resulting in more unpredictable yields. Federal crop insurance or commodity program provisions that exempt double-cropped fields from program participation if farms are in areas where the chosen double-cropped combinations are not widely accepted as a viable practice by local agricultural experts may exacerbate these barriers to double-cropping adoption. Much of the information on whether or whether farmers double crop (and, if so, how often) is based on anecdotal testimony. Climate variables likely contribute to variances in double cropping, with longer growing seasons providing more opportunity for double cropping. Similarly, variations in annual weather may have an impact on year-to-year variances. Warmer winters or springs, for example, allow more farmers in each region to plant a spring crop after their winter ones. Commodity prices may also be a factor, as higher prices provide greater incentives to double crop.

PERMACULTURE

Introduction

Permaculture is a concept and approach to land management that is based on the arrangements seen in healthy natural ecosystems. It consists of a set of design ideas drawn from whole-systems thinking. It applies these principles to regenerative agriculture, rewilding, and community resilience, among other things. Permaculture was derived from the term “permanent agriculture,” but it was later expanded to include social components, as influenced by Masanobu Fukuoka’s natural farming. Bill Mollison and David Holmgren created the term in 1978 to contrast Western industrialized approaches with Indigenous or traditional knowledge. Ecological design, ecological engineering, regenerative design, environmental design, and building are all disciplines of permaculture. It also incorporates regenerative and self-maintained habitat and agricultural systems fashioned after natural ecosystems, as well as integrated water resources management that generates sustainable architecture. Permaculture has been widely adopted as an agricultural and architectural design approach, as well as a guiding life principle or philosophy, all around the world. The role of Indigenous knowledge and traditions, on which the practice is rooted, has been credited with much of its success. As a result of the emergence of permaculture, Indigenous knowledge has being revalidated in circles where it had previously been dismissed.

Suburban and Urban Permaculture

The optimal use of space is a vital component in suburban and urban permaculture. It’s critical to maximize space for food production while minimizing waste. To overcome this issue of space, the Wildfire magazine advises utilizing approaches like the keyhole garden. Neighbors can also work together to expand the scope of the transformation. Recreation facilities, neighborhood associations, city programs, church groups, and schools are all examples of places that might become part of a bigger social and economic movement. Columbia, an ecovillage of 37 apartment condominiums in Portland, Oregon, pushed its neighbors to adopt similar green-minded principles or permaculture, such as front-yard gardens. Rainwater catchment, edible planting, removing asphalt drives, converting a garage into living space, converting a south side patio to passive solar, aesthetic features, and detached structures are all included in suburban permaculture sites like one in Eugene, Oregon. Creating community-managed agricultural or farm sites involves transforming vacant areas in suburban and urban contexts. Those in power, on the other hand, see some of these farm sites as temporary or informal solutions to vacant lots rather than permanent municipal fixtures. This puts the permaculture principle of permanence under jeopardy. For example, despite widespread protest from the majority Latino community who had developed deep bonds with the site, Los Angeles' South Central Farm (1994-2006) was bulldozed with approval from property owner Ralph Horowitz, despite large-scale protest from the majority Latino community who had developed deep bonds with the site. Approximately 40 farmers have been arrested and evicted. The land sat empty for over a decade before being approved for offices and warehouses by the city council in 2019. The possibilities and challenges for developing suburban or urban permaculture differ greatly as a result of how the built environment is designed and property is treated in different parts of the world. A research comparing the built environment of Jaisalmer, India, with Los Angeles, California, found that the American planned metropolis is ecologically disastrous:

Design Principles

- **Observe and interact:** Take time to engage with nature to design solutions that suit a particular situation.
- **Catch and store energy:** Develop systems that collect resources at peak abundance for use in times of need.
- **Obtain a yield:** Emphasize projects that generate meaningful rewards.
- **Apply self-regulation and accept feedback:** Discourage inappropriate activity to ensure that systems function well.
- **Use and value renewable resources and services:** Make the best use of nature's abundance: reduce consumption and dependence on non-renewable resources.
- **Produce no waste:** Value and employ all available resources: waste nothing.
- **Design from patterns to details:** Observe patterns in nature and society and use them to inform designs, later adding details.
- **Integrate rather than segregate:** Proper designs allow relationships to develop between design elements, allowing them to work together to support each other.
- **Use small and slow solutions:** Small and slow systems are easier to maintain, make better use of local resources and produce more sustainable outcomes.
- **Use and value diversity:** Diversity reduces system-level vulnerability to threats and fully exploits its environment.
- **Use edges and value the marginal:** The border between things

is where the most interesting events take place. These are often the system's most valuable, diverse, and productive elements.

- **Creatively use and respond to change:** A positive impact on inevitable change comes from careful observation, followed by well-timed intervention.

Foundational Ethics

The ethics on which permaculture builds are:

- **Earth care:** Provision for all life systems to continue and multiply.
- **People care:** Provision for people to access those resources necessary for their existence.
- **Fair share:** Setting limits to population and consumption so that people do not take more than what is needed. By governing our own needs, we can set resources aside to further the above principles. This principle is also described as share the surplus.

Landscape, function, and species assembly patterns are all emphasized in permaculture. It decides where these elements should be placed in order to benefit the local ecosystem the most. Permaculture enhances the value of links between components and the final design's synergy. As a result, the focus of permaculture is on the interactions between the elements rather than on the individual elements themselves. When done correctly, the total is larger than the sum of its parts. Permaculture aims to maximize benefits while minimizing waste, human work, and energy input. Because these solutions have evolved over thousands of years and have proven to be effective, permaculture design is based on reproducing or imitating natural patterns seen in ecosystems. As a result, permaculture design implementation will vary greatly depending on the location of the Earth in which it is implemented. The science of systems ecology and the study of pre-industrial instances of sustainable land use have influenced design concepts. Organic farming, agroforestry, integrated farming, sustainable development, physics, meteorology, sociology, anthropology, biochemistry, engineering, and applied ecology are all used in permaculture.

Advantages of Permaculture

Protecting the planet's resources: The major ways to accomplish this are to use natural forms of energy (solar, wind), save resources such as water (for example, by collecting rainwater), and avoid using fossil fuels. We can also make the most of the natural resources available to us. This includes things like burning deadwood for heat or growing mushrooms, harvesting medicinal plants that grow naturally in the wild, growing anti-pollution plants to improve indoor air quality, strategically planting trees to provide shade in the summer, and keeping chickens and ducks to keep pests at bay. We must also avoid all forms of trash and ensure that waste items are reused.

- **Less work and energy required:** In a vegetable garden, one permaculture strategy is to avoid actively cultivating the soil. Instead, the soil should be allowed to be naturally digested by insects (with a little encouragement from some natural mulch). Underground insect populations can be severely harmed by using a tiller or cultivator.
- **Promoting biodiversity and preserving wildlife:** Permaculture, unlike traditional agriculture, does not work against nature. It, on the other hand, draws inspiration from nature and collaborates with it, celebrating and profiting from the great diversity it offers.
- **Organic produce:** In order to maintain biodiversity, permaculture avoids using herbicides and insecticides, even those of organic origin. Herbicides and artificial fertilizers are never utilized because nature is expected to self-regulate. Instead of using insecticides, a permaculture gardener can choose to surrender a percentage of their produce to pests.
- **Boosting health:** Permaculture is beneficial to your health not only because you'll be eating organic vegetables, but also because you'll be spending more time on yourself and others. Spending more time with your children, family, and loved ones is a great way to combat stress and other factors that might harm our mental health.
- **Saving space:** Producing a great number of produces in a short location takes up less space, allowing more room for other activities. Finally, there is one benefit that should not be overlooked: you will have the satisfaction of knowing that you are contributing to the larger good while doing so with minimal environmental impact.

FOREST GARDENING

Introduction

Forest gardening is a low-maintenance, sustainable plant-based food production and agroforestry system based on woodland ecosystems that includes fruit and nut trees, shrubs, herbs, vines, and perennial vegetables with outputs that are directly useful to humans. These can be intermixed to develop in a succession of layers to create a woodland habitat using companion planting. In tropical climates, forest gardening is a primitive means of procuring food. After modifying the techniques and applying them to temperate settings in the 1980s, Robert Hart coined the phrase "forest gardening." Hart started farming in Shropshire's Wenlock Edge with the goal of producing a healthy and therapeutic environment for himself and his brother Lacon. Starting out as traditional smallholders, Hart quickly learned that maintaining huge annual vegetable beds, producing animals, and caring for an orchard were all jobs beyond their capabilities. A tiny bed of perennial vegetables and herbs he planted, on the other hand, was taking care of itself with minimal help. Hart replaced his farm animals with plants after adopting a raw vegan diet for health and personal reasons. Fruit, nuts, and green leafy vegetables are the three major products of a forest garden. He planned to call his gardening method ecological horticulture or eco-cultivation after a model woodland garden he constructed from a 0.12-acre (500 m2) orchard on his estate. When Hart saw that agroforestry and forest gardens were already being used to describe similar systems in other regions of the world, he discarded these labels. Toyohiko Kagawa and James Sholto Douglas' Forest gardening methods, as well as the productivity of Keralan home gardens, inspired him, as Hart explains: "The Indian state of Kerala, which has over three and a half million forest gardens, is perhaps the world's most advanced country in terms of agroforestry... A study group discovered twenty-three young coconut palms, twelve cloves, fifty-six bananas, and forty-nine pineapples on a plot of only 0.12 hectares (0.30 acres), with thirty pepper vines trained up its trees, as an illustration of the exceptional intensity of cultivation of some forest gardens. The little landowner also cultivated fodder for his house-cow."

Seven-Layer System

Robert Hart developed a system based on the idea that the natural forest can be split into tiers. He employed intercropping to transform a modest apple and pear orchard into an edible polyculture landscape with the following layers:

- 'Canopy layer' consisting of the original mature fruit trees.
- 'Low-tree layer' of smaller nut and fruit trees on dwarfing root stocks.
- 'Shrub layer' of fruit bushes such as currants and berries.
- 'Herbaceous layer' of perennial vegetables and herbs.
- 'Rhizosphere' or 'underground' dimension of plants grown for their roots and tubers.
- 'Ground cover layer' of edible plants that spread horizontally.
- 'Vertical layer' of vines and climbers.

The plants he chose were an important part of the seven-layer structure. Most today's typical vegetable crops, such as carrots, are sun-loving plants that aren't well suited to the shadier forest garden environment. Shade-tolerant perennial vegetables were Hart's favorites.

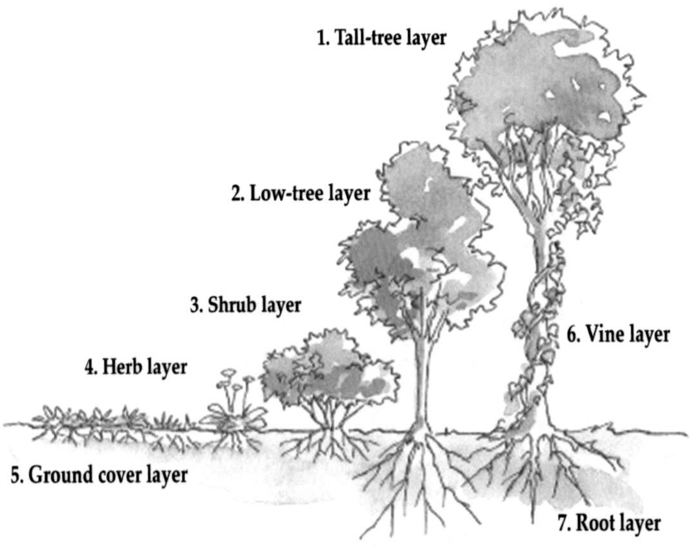

Figure 26: Seven-layer system.

Source: https://nationalforestgardening.org/benefits-of-forest-gardens/key-benefits/

The Benefits of Forest Gardening

Forest gardens are grown for a variety of reasons, and while most intense annual vegetable plots are quite similar, each forest garden is unique because it is created to meet the needs and expectations of its users. Some of the reasons why forest gardens are cultivated are as follows.

- **Working with the land instead of against it:** The climax vegetation in a moist temperate environment is woodland or forest - that is, if you leave a piece of land alone, it will eventually become a forest; the forces of nature are actively driving the area towards woodland. The further your agricultural or horticultural system is from woodland, the more energy it needs to maintain it, and the more disturbed and removed it is from a long-term biological state. As a result, arable fields or annually cultivated land consume the greatest energy; pasture consumes less; and orchard systems consume even less. Natural woods require no human effort to maintain because it takes care of itself. Forest gardens are a hybrid of orchards and natural woodland, and they provide some of the most beneficial goods with the least amount of energy input.
- **Low maintenance and high, efficiency:** A forest garden will include a variety of trees, shrubs, perennials, and annuals. It can probably contain annual vegetables, but most of the ground is covered with herbaceous perennial plants that are used directly or indirectly. Apart from occasional pruning, trees and shrubs require little care. One of the most important aspects of forest gardening is to attempt to maintain most of the soil always covered with plant growth or plant matter; this keeps the soil healthy, which benefits all the other plants. Weeding consumes a major portion of the maintenance time in annual plant gardens; however, weeding is minor in forest gardens since there is little space or chance for weeds to establish. There are usually few gaps between plants in a forest garden, and one plant may need to be pruned back to make place for others, whereas in 'conventional' gardening, plants are spaced with bare ground beneath them, which is perfect for weed infestations and certain to increase care time. Any agricultural system's biological efficiency is defined as the ratio of energy outputs to energy inputs; it is not the same as output or yield. Forest gardens are highly efficient since they are

low-input systems. Depending on the design, the outputs might range from low to high. Consider an apple orchard. Tree-based systems can probably produce as much as arable lands.

- **Wide range of products:** Forest gardens can provide a wide range of products, such as fruits, nuts, seeds, vegetables, salad crops, herbs, spices, firewood, mushrooms grown on logs, poles, and canes, tying materials, basketry materials, medicinal herbs, dye plants, soap plants, honey from bees, sap products, and so on.
- **High nutritional value:** There is a lot of evidence suggesting crops grown on permanent plants are more nutritious than crops grown on annual plants. Perennial plants' root systems are more extensive and perennial, allowing them to exploit soil space more efficiently than annual plants and hence accumulate higher amounts of nutrients.
- **Resilience to climate extremes:** In the event of harsh weather, forest-based systems are the most durable. A forest garden's structure and diversity ensure strong resilience, for example, to the effects of climate change, which will include more harsh weather events.
- **Biologically sustainable:** The diversity of forest gardens, as well as the complex network of below- and aboveground interactions between species, ensures their long-term viability. Forest gardens have been around for over a thousand years in the tropics. Because of their intricacy, agricultural scientists rarely examine them; modeling two species growing together is tough enough; modeling the 100 or 200 species that most forest gardens contain is far beyond reductionist science approaches.
- **Aesthetically beautiful:** Whether aesthetic goals were part of the design process, forest gardens are quite lovely settings. A forest garden does not feel like a 'regular' manicured garden; it feels wilder, more jungle-like in places, less maintained, and less interfered with. Forest gardens can link people to a bountiful nature in a way that excursions to nature reserves cannot, because we are all participants in nature and consumers of the food and other things that nature supplies. Although most plants used in forest gardens offer direct or indirect human benefits, they can also be ornamental, and plants can be added only for ornamental purposes if desired.

- **Environmentally beneficial:** Forest gardens provide numerous environmental advantages. Carbon dioxide is sequestered in the soil and in the woody biomass of trees and shrubs. Emissions of greenhouse gases are insignificant. Forest gardens are excellent at holding water after heavy rains and minimizing flooding and erosion by keeping the soil covered and in good shape. They can provide cover for structures, reducing the amount of energy used for heating. They're also great for wildlife (some of which may turn into invasive species!); the intricate three-dimensional structure and diversity of plants (both native and non-native) provide a variety of niches for insects and tiny animals. A recent invertebrate study in our own forest garden discovered that the forest garden had a higher diversity of species than a planted native woodland of the same age, possibly indicating that plant species diversity is more important than their origin.
- **Commercial potential:** Even if it's just one or two crops from fruit trees, such as a bumper crop of apples or mangoes, most forest gardens around the world contain a commercial component. Certain features should be introduced during the design stage (see Chapter 9, page 92) to exploit a forest garden on a more serious commercial scale, particularly a limit to the diversity (for example, 300 species would be too complex to manage efficiently) and changes to the positioning of species to make harvesting more efficient.

AQUAPONICS

Introduction

Aquaponics is a food production system that combines aquaculture (raising aquatic animals such as fish, crayfish, snails, or prawns in tanks) and hydroponics (cultivating plants in water), in which nutrient-rich aquaculture water is fed to hydroponic-grown plants, with nitrifying bacteria converting ammonia to nitrates. Because all aquaponic systems are based on current hydroponic and aquaculture farming techniques, the size, complexity, and types of foods grown in an aquaponic system can vary as much as any system found in either distinct agricultural discipline. Aquaponics has a long history, though there is some controversy about when it initially appeared:

- Aztec cultivated agricultural islands known as chinampas in a system considered by some to be an early form of aquaponics for agricultural use, where plants were raised on stationary (or sometime movable) islands in lake shallows and waste materials dredged from the Chinampa canals and surrounding cities were used to manually irrigate the plants.
- South China and the whole of Southeast Asia, where rice was cultivated and farmed in paddy fields in combination with fish, are cited as examples of early aquaponics systems, although the technology had been brought by Chinese settlers who had migrated from Yunnan around 5 AD. These polyculture farming systems existed in many Far Eastern countries and raised fish such as the oriental loach, swamp eel, common carp and crucian carp as well as pond snails in the paddies.
- The 13th century Chinese agricultural manual Wang Zhen's Book on Farming described floating wooden rafts which were piled with mud and dirt and which were used for growing rice, wild rice, and fodder. Such floating planters were employed in regions constituting the modern provinces of Jiangsu, Zhejiang, and Fujian. These floating planters are known as either jiatian or fengtian, which translates to "framed paddy" and "brassica paddy", respectively. The agricultural work also references earlier Chinese texts, which indicated that floating raft rice cultivation was being used as early as the Tang Dynasty (6th century) and Northern Song Dynasty (8th century) periods of Chinese history.

In recent years, China has erected large-scale floating aquaponics systems on polyculture fish ponds. Rice, wheat, canna lilies, and other crops are grown on them, with some installations covering more than 2.5 acres (10,000 m2). The New Alchemy Institute's various works and the work of Dr. Mark McMurtry et al. at North Carolina State University, who devised an Integrated-Aqua-Vegeculture-System (iAVs) based on the combination of aquaculture and sand-based grow beds, are often credited with the development of modern aquaponics. Other institutes quickly followed in the footsteps of the New Alchemy Institute and McMurtry's iAVs. Dr. James Rakocy and his University of the Virgin Islands colleagues began researching and developing the use of deep-water culture hydroponic grow beds in a large-scale aquaponics system in 1979. Other institutes focused their research on "ebb and flow" (also known as "flood and drain") systems, which were partially based on the original ideas developed at North Carolina

State University, but where coarse media (such as gravel or expanded clay) replaced sand and bell syphons allowed an ebb-and-flow irrigation cycle. Such systems are also known as "Speraneo Systems" because they are based on the original ideas developed at North Carolina State University. The first aquaponics research in Canada took place at a research station in Lethbridge, Alberta, when a tiny system was added to existing aquaculture research. Throughout the 1990s, Canada saw an increase in aquaponics setups, mostly as commercial installations cultivating high-value crops like trout and lettuce. In a greenhouse in Brooks, Alberta, where Dr. Nick Savidov and colleagues researched aquaponics from a plant science background, a setup based on the deepwater system developed at the University of Virgin Islands was built. The researchers discovered rapid root growth in aquaponics systems and closed the solid-waste loop, and discovered that the system can run effectively at a low pH level, which is preferred by plants but not by fish, due to certain benefits in the system over traditional aquaculture.

Parts of an Aquaponic System

Aquaponics is divided into two parts: aquaculture and hydroponics. Aquaculture is used to raise aquatic animals, while hydroponics is used to cultivate plants. Due to the closed-system recirculation of most aquaculture systems, aquatic effluents from uneaten feed or raising animals like fish collect in water. In high proportions, the effluent-rich water is hazardous to aquatic animals, yet it includes nutrients necessary for plant growth. Even though aquaponics systems are essentially made up of these two portions, they are frequently divided into multiple components or subsystems that are responsible for effective solid waste disposal, supplying bases to neutralize acids, or sustaining water oxygenation. The following are examples of typical components:

- **Rearing tank:** the tanks for raising and feeding the fish;
- **Settling basin:** a unit for catching uneaten food and detached biofilms, and for settling out fine particulates;
- **Biofilter:** a place where the nitrification bacteria can grow and convert ammonia into nitrates, which are usable by the plants;
- **Hydroponics subsystem:** the portion of the system where plants are grown by absorbing excess nutrients from the water;
- **Sump:** the lowest point in the system where the water flows to and from which it is pumped back to the rearing tanks.

The units for solids removal, biofiltration, and/or the hydroponics subsystem may be merged into one unit or subsystem, preventing water from moving straight from the aquaculture part of the system to the hydroponics part, depending on the sophistication and expense of the aquaponics system. Solids are collected and the medium has enough surface area for fixed-film nitrification when gravel or sand is used as a plant supporting media. The ability to combine biofiltration and hydroponics allows aquaponic systems to eliminate the requirement for a separate, expensive biofilter in many circumstances.

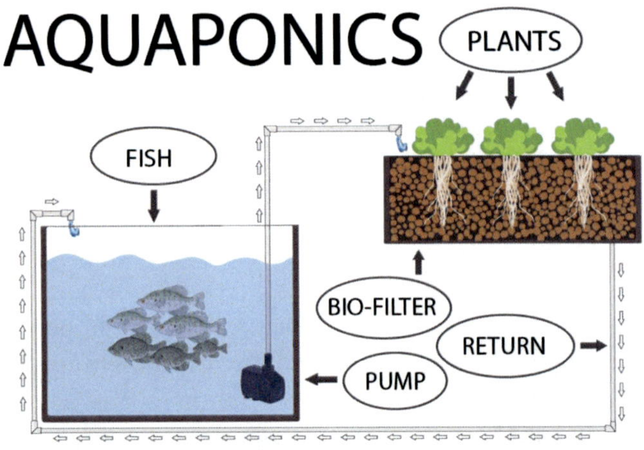

Figure 27: Aquaponic system.

Source: Google Images

Live Components

To function properly, an aquaponic system relies on a variety of live components. Plants, fish (or other aquatic species), and microbes are the three main living components. Additional live components, such as worms, are included in certain systems.

- **Plants:** Plants grow directly into effluent-rich water without the use of a soil medium in a Deep-Water Culture hydroponics system. Because the roots do not need to grow outwards to sustain the plant's weight, plants can be spaced closer together. In a Nutrient Film Technique (NFT) system, a plant is placed in a nutrient-rich water channel. Many plants are suited for aquaponic systems, those that function best for a given system are determined by the fish's maturity and stocking density. These parameters have an impact on the concentration of nutrients in fish waste and how much of that nutrient is made available to plant roots via bacteria. Green green vegetables with low to medium nutrient requirements, such as Chinese cabbage, lettuce, basil, spinach, chives, herbs, and watercress, are ideally suited to aquaponic systems. Other plants with higher nutrient requirements, such as tomatoes, cucumbers, and peppers, will thrive only in mature aquaponic systems with high fish stocking densities. Cucumbers, shallots, tomatoes, lettuce, capsicum, red salad onions, and snow peas are just a few of the salad plants that have thrived in aquaponics. Chinese cabbage, lettuce, basil, roses, tomatoes, okra, cantaloupe, and bell peppers are all profitable plants for aquaponic systems. Watercress, basil, coriander, parsley, lemongrass, sage, beans, peas, kohlrabi, taro, radishes, strawberries, melons, onions, turnips, parsnips, sweet potato, cauliflower, cabbage, broccoli, and eggplant, as well as the choys used in stir fries, all grow well in an aquaponic system.

- **Fish (or other aquatic creatures):** Filtered water from the hydroponics system is recirculated in a catfish tank. Due to their capacity to endure crowding, freshwater fish are the most popular aquatic animal farmed in aquaponics, while freshwater crayfish and prawns are also occasionally utilized. Saltwater aquaponics is a type of aquaponics that uses saltwater fish. Many warmwater and coldwater fish species have adapted well to aquaculture systems. Because it is a warmwater fish species that can handle crowding and changing water conditions, tilapia is the most preferred fish for home and commercial projects aimed at raising edible fish. Also utilized are barramundi, silver perch, eel-tailed catfish (tandans catfish), jade perch, and Murray cod. Bluegill and catfish are ideal fish species for home systems in temperate locations where the ability or willingness to manage water temperature is lacking. If the fish in the system do not need to be edible, koi and goldfish can be employed. Channel catfish, rainbow trout, perch, common carp, Arctic char, largemouth bass, and striped bass are all good choices.
- **Bacteria:** Nitrification, or the aerobic conversion of ammonia to nitrates, is one of the most significant activities in an aquaponic system since it lowers the toxicity of the water for the fish

and allows the plants to use the generated nitrate compounds for nutrition. As a byproduct of fish metabolism, ammonia is continuously discharged into the water through their excreta and gills. However, greater quantities of ammonia (usually between 0.5 and 1 ppm) can hamper growth, cause widespread tissue damage, reduce disease resistance, and even kill the fish. Although plants can absorb ammonia from the water to some extent, nitrates are more easily digested, decreasing the water's toxicity for fish. Nitrosomonas, which converts ammonia to nitrites, and Nitrobacter, which converts nitrites to nitrates, can be used to convert ammonia into safer nitrogenous chemicals. While nitrite is still damaging to fish because it can produce methemoglobin, which is unable to bind oxygen, nitrates can be tolerated at high levels by fish because they bond to hemoglobin. The higher the surface area, the more room there is for nitrifying bacteria to flourish. Material selection for grow beds necessitates careful consideration of surface area, cost, and maintainability.

Hydroponic Subsystem

Plants are cultivated in the same way that they are in hydroponics, with their roots immersed in nutrient-rich effluent water. This allows them to filter out ammonia or its metabolites, which are hazardous to aquatic animals. The water is cleansed and oxygenated after passing through the hydroponic subsystem, and it can then be returned to the aquaculture vessels. This cycle continues indefinitely. The following are some of the most common aquaponic applications of hydroponic systems:

- **Deep-water raft aquaponics:** In troughs, styrofoam rafts float in a moderately deep aquaculture basin. Raft tanks can be built to be rather big, allowing seedlings to be transplanted at one end of the tank while fully grown plants are harvested at the other, allowing for the most efficient use of floor area.
- **Recirculating aquaponics:** solid medium, like as gravel or clay beads, held in a container inundated with aquaculture water. Closed-loop aquaponics is another name for this sort of aquaponics.
- **Reciprocating aquaponics:** solid media in a container with multiple sorts of siphon drains that are alternately inundated and drained. Flood-and-drain aquaponics or ebb-and-flow aquaponics are other names for this form of aquaponics.

- **Nutrient film technique channels:** Plants are grown in long, narrow channels with a continuous film of nutrient-rich water running through the roots. Because the little volume of water and narrow passageways make it impossible for beneficial bacteria to thrive, this method necessitates the use of a bio filter.
- Other systems make use of top-fed towers, horizontal PVC pipes with holes for the pots, and half-filled plastic barrels with pebbles or rafts. Each method has its own set of advantages.

Plant harvesting is staggered, with seedlings developing at the same time as mature plants, because different growth stages require varying amounts of minerals and nutrients. Because of the continual symbiotic cleansing of pollutants from the water, the nutrient content of the water remains consistent.

Biofilter

The bacteria that convert ammonia to useable nitrates for plants in an aquaponics system build a biofilm on any solid surfaces in the system that are in continual contact with the water. The total surface area of the submerged roots of the vegetables provides a vast surface area where many germs can accumulate. The speed of nitrification is determined by the surface area of the water, as well as the quantities of ammonia and nitrites in the water. It's critical to look after these bacterial colonies in order to ensure that ammonia and nitrite are properly assimilated. Therefore, in most aquaponics systems, a biofiltering device is included to aid in the growth of these bacteria. Ammonia levels typically range from 0.25 to.50 ppm after a system has stabilized; nitrite levels vary from 0.0 to 0.25 ppm; and nitrate levels range from 5 to 150 ppm. Ammonia (up to 6.0 ppm) and nitrite (up to 15 ppm) levels may increase during system starting, with nitrate levels peaking later in the process. Ammonia is converted to nitrite, which releases hydrogen ions into the water during the nitrification process. A person's pH will gradually drop over time, therefore they can utilize non-sodium bases like potassium hydroxide or calcium hydroxide to neutralize the pH of the water if there are insufficient amounts naturally present to offer a buffer against acidification. Additionally, in addition to the fish waste that serves as the primary source of nutrients for plants, specific minerals, or nutrients such as iron can be provided. Worms, which liquefy solid organic matter so that it may be used by the plants and/or other animals in the system, are a useful technique to deal with solids buildup in aquaponics. Please check Vermiponics for a worm-only growth method.

Operation

Water, oxygen, light, nutrition for aquatic animals, and electricity to pump, filter, and oxygenate the water are the system's five key inputs. To keep the system stable, spawn or fry can be added to replace grown fish that have been removed. In terms of outputs, an aquaponics system can produce plants such as hydroponic vegetables and edible aquatic animals raised in an aquaculture system on a continuous basis. For every 1 US gal (3.8 L) of aquaculture water in the system, a typical build ratio is.5 to 1 square foot of grow space. Depending on aeration and filtration, 1 US gal (3.8 L) of water can support between.5 lb (0.23 kilogram) and 1 lb (0.45 kg) of fish stock. Dr. James Rakocy, the director of the University of the Virgin Islands' aquaponics research team, announced ten fundamental guiding principles for developing successful aquaponics systems, based on substantial research conducted as part of the Agricultural Experiment Station aquaculture program.

- Use a feeding rate ratio for design calculations
- Keep feed input relatively constant
- Supplement with calcium, potassium, and iron
- Ensure good aeration
- Remove solids
- Be careful with aggregates
- Oversize pipes
- Use biological pest control
- Ensure adequate biofiltration
- Control pH

Feed source: Stock feed is frequently made up of fish meal obtained from lower-value species, as it is in most aquaculture-based systems. This technique is unsustainable due to the ongoing decline of wild fish stocks. Organic fish meals could be a potential alternative to alleviate this worry. Other options include producing duckweed in an aquaponics system that feeds the same fish, excess worms from vermiculture composting, using prepared kitchen scraps, and cultivating black soldier fly larvae in composting grub growers to feed to the fish.

Plant nutrients: A few minerals and micronutrients, like hydroponics, can be added to boost plant development. In aquaponics, iron is the most lacking nutrient, but it can be supplemented by mixing Iron Chelate powder with water. Potassium sulfate, a foliar spray, can be used to supplement potassium. Magnesium, in the form of Epsom salt, calcium, in the form of calcium chloride, and boron are less important nutrients. Aquaculture wastes undergo biological filtration, which results in high nitrate concentrations, which are beneficial to leafy greens. Supplemental nutrients such as magnesium, calcium, potassium, and phosphorus are advised for blooming plants with high nutrient demands. Potash sulfate, potassium bicarbonate, monoammonium phosphate, and other common sources are used. Raw or mineralized sludge, which typically has 3–17 times greater nutritional contents, can entirely offset nutrient shortage in wastewater from fish component (RAS). RAS effluents (wastewater and sludge combined) contain enough nitrogen, phosphorus, magnesium, calcium, sulfur, iron, zinc, copper, and nickel to suit the needs of most aquaponic crops. Potassium deficiency is common, necessitating full-fledged fertilization. Micronutrients B and Mo are insufficient in some cases, although this can be easily remedied by increasing sludge release. The assumption that 'definite' phytotoxic salt levels in RAS effluents exist should be challenged; there are also practical solutions. Within the aquaponics loop, there is no danger of heavy metal accumulation.

Water usage: In normal operation, aquaponic systems do not discharge or exchange water, but instead recirculate and reuse it very effectively. To maintain a stable aquatic environment with little variation in ambient nutrient and oxygen levels, the system relies on the relationship between the animals and the plants. Plants can recover dissolved nutrients from circulating water, resulting in less water being discharged and a lower water exchange rate. Water is only given to replace water lost by plant absorption and transpiration, surface water evaporation into the air, system overflow from rainfall, and biomass removal such as settled solid wastes. As a result, aquaponics consumes around 2% of the water that a traditionally irrigated farm would use to produce the same quantity of vegetables. This enables aquaponic agricultural and fish cultivation in regions where water and fertile soil are scarce. Controlled wetland conditions can also be replicated using aquaponic systems. Biofiltration and treatment of average residential sewage can be accomplished using constructed wetlands. The nutrient-rich overflow water can be collected in catchment tanks and used to boost the growth of soil-grown crops, or it can be piped back into the aquaponic system to replenish the water supply.

Energy usage: To accomplish recirculation and water/ambient temperatures, aquaponic installations rely to varied degrees on man-made energy, technological solutions, and environmental management. However, if a system is constructed with energy saving in mind, using alternative energy and fewer pumps by allowing water to flow downwards as much as possible, it can be extremely energy efficient. While careful design can reduce the danger, aquaponics systems can have several "single points of failure," where issues like an electrical failure or a pipe obstruction might result in the loss of the entire fish stock.

Fish stocking

The hydroponic plant components and fish rearing components must nearly always be at maximum production capacity for aquaponic systems to be financially successful and create a profit while also covering running expenditures. There are three basic stocking methods that can assist keep the bio-mass of fish in the system at its maximum (without inhibiting fish growth).

- **Sequential rearing:** Multiple age groups of fish share a rearing tank, and when one reaches market size, it is harvested and replaced with the same number of fingerlings. The disadvantages of this strategy include stressing out the entire pool of fish during each harvest, missing fish, which wastes food and space, and the difficulty of keeping correct records with frequent harvests.
- **Stock splitting:** When the tank reaches capacity, large numbers of fingerlings are introduced at once and then separated into two groups, making it easier to keep track of and preventing fish from being "lost." The use of "swim-ways," which connect several rearing tanks and a series of hatches/moving screens/pumps that move the fish around, is a stress-free approach to accomplish this procedure.
- **Multiple rearing units:** Once their current tank reaches capacity, entire groups of fish are transported to larger rearing tanks. When the largest tank is harvested, the other fish groups are each transferred up into a bigger tank, while the smallest tank is refilled with fingerlings in such systems. It's also usual to have many rearing tanks but no method to transport the fish between them, which eliminates the work of moving the fish and allows each tank to be unaffected during harvesting, even though the space consumption is wasteful when the fish are fingerlings.

In order to avoid stress from overcrowding, properly feed the fish, and promote healthy growth, the bio-mass of fish in the rearing tanks should not exceed 0.5 lbs/gallon.

Disease and Pest Management

Pesticides are generally used to control insects on crops, but in an aquaponic system, pesticides would endanger the fish environment. Medications, on the other hand, cannot be utilized if the fish contract parasites or diseases since the plants would absorb them. Non-chemical pest control strategies such as traps, physical barriers, and biological control (such as parasitic wasps/ladybugs to control white flies/aphids) should be employed to maintain the symbiotic interaction between the plants and the fish. Neem oil is the most effective organic insecticide; however, it should only be used in small amounts to avoid spilling into fish water. Pest and disease management bottlenecks frequently stymie aquaponics commercialization. Chemical control approaches are extremely difficult to utilize in any system. While well-established commercial biocontrol techniques can replace insecticides and herbicides, fungicides and nematicides are still useful in aquaponics. The initial steps in controlling insect populations are monitoring and cultural control. Biological controls, on the other hand, are more adaptive. In all designs, nonchemical prophylactic techniques are highly effective for pest and disease prevention.

Automation, monitoring, and control

Many people have attempted to develop automatic control and monitoring systems, and some have had some success. Researchers, for example, were able to automate a small-scale aquaponic system to create a cost-effective and long-term agricultural method. Commercial automation technology development has also emerged. For example, a business has created a system that can automate farming's repetitive duties and includes a machine learning algorithm that can detect and eradicate unhealthy or underdeveloped plants automatically. The automated technology is part of a 3.75-acre aquaponics plant that claims to be the first indoor salmon farm in the United States. The aquaponic machine has made significant progress in terms of aquaponics documentation and data collection.

Economic Viability

Aquaponics is a versatile and stable polyculture technology that allows farmers to simultaneously cultivate crops and breed fish. Farmers can continue to make money even if the market for fish or vegetables is in a slump because they have two streams of income. An aquaponic system's versatility allows it to grow a wide range of products, including common vegetables, herbs, flowers, and aquatic plants, to appeal to a wide range of customers. Due to their modest nutritional requirements, herbs, lettuce, and specialist greens like basil or spinach are especially well suited for aquaponic systems. Aquaponic goods are organic and pesticide-free, with a little environmental footprint, which appeals to an increasing number of ecologically aware consumers. Aquaponic systems are also cost-effective because to low water usage, effective nutrient cycling, and the fact that they require little acreage to operate. Aquaponic systems can be put up in locations with typically poor soil quality or contaminated water because they don't require soil and only a little amount of water. More importantly, aquaponic systems are usually free of weeds, bugs, and illnesses that afflict soil, allowing them to produce high-quality crops reliably and fast.

Conservation Tillage

Any type of soil cultivation that leaves the previous year's crop residue (such as corn stalks or wheat stubble) on fields before and after planting the next crop to prevent soil erosion and runoff, as well as other benefits like carbon sequestration, is known as conservation tillage (MDA, 2011). Following planting, at least 30% of the soil surface is covered with crop residue/organic residue using this procedure. It also has dirt that is not inverted. Tillage depth and the proportion of surface area disturbed define this form of soil tillage. For example, in order to include Treflan, which was sprayed in front of the machine, the planter was modified to place the seed 50mm deep and give a coating of fine tilth 18mm deep across the planted row regions. All of this happened at a speed of 20 km/h. Zero-till, strip-till, ridge-till, and mulch-till are examples of conservation tillage techniques. Zero-tillage is a type of conservation tillage that causes the least amount of disturbance to the soil surface. Planting crops directly into residue that hasn't been tilled at all is known as zero-till (MDA, 2011). Because huge machineries are required for planting, zero tillage technology is commonly used in large-scale agricultural crop production systems. There is no suitable machinery for sowing on smaller farms, though extremely small-scale farmers can do so by hand. Planting crops with zero-tillage causes the least amount of soil disturbance by sowing seeds in an un-ploughed field with no extra land preparation. In one operation, a typical zero-tillage machine can spread seed in incisions 2-3cm wide and 4-7cm deep, as well as apply fertilizer (CIMMYT, 2010). To open the slits, the machine has an inverted T-type furrow opener. Seed and fertilizer are automatically placed in the appropriate boxes and dropped into the slits. A hydraulic device on the tractor may regulate the depth of the slots.

Conservation Tillage Practices

- Soil management is an important part of a long-term agricultural production system.
- In a conservation tillage system, maintaining an effective and acceptable amount of residue is crucial.
- In a conservative tillage technique, managing soil erosion, preserving appropriate organic matter, nutrient delivery, and soil moisture for crop productivity are all important elements to consider.
- Conservation tillage technologies have advantages and downsides that must be evaluated depending on the land environment.

Advantages:

- 60 to 90 percent residue stays on the soil surface, depending on tillage type, to assist decrease erosion. Soil particles can be protected from rain and wind erosion by residue.
- Reduces evaporation at the soil surface, so conserving soil moisture.
- Increases organic matter, which helps to enhance soil structure and boost water infiltration, and so improves soil health and lowers runoff.
- Because there are fewer tillage passes across the field, it requires less labor and fuel and lowers soil compaction.
- During harvest, the earth is firmer.
- Dust and diesel pollutants may cause less air pollution.
- Wildlife can find food and shelter in the residue.

Disadvantages:

- In the spring, slow soil warming is important, especially on poorly drained soils.
- Planter modification or a no-till planter are required.
- Increased reliance on chemicals to control weeds.
- Because insect, disease, and weed concerns may differ from pests encountered in traditional tillage systems, scouting is essential.

Conservation Tillage Systems

- **No-till:** Crop residue on the soil surface is not disturbed in this technique, and it should remain on the surface from harvest to seeding, and from seeding to harvest. This approach aids in soil erosion reduction, particularly on highly erodible soils, as well as appropriate stand establishment. The soil is only disturbed when a coulter, seed furrow opener, row cleaner, or other attachments to planters or drills form a narrow zone for planting or drilling operations. Cutting the crop residue and penetrating the undisturbed soil should be possible with no-till planters and drills. Depending on soil types, field conditions, and residue source, between 60 to 70% of residue is required after planting to manage erosion and conserve soil moisture. The system's drawbacks include reliance on herbicides for weed control, the need for a no-till planter or planter modification, the possibility of planting delays due to damp, chilly soil temperatures, and issues with various insect, disease, and weed species.
- **Strip-till:** It's typically employed in fields that are poorly drained or have a low slope. This technique is most useful when the weather is cold and damp in the spring. Strip-tilling is like no-tilling in that narrow strips of the field are tilled while the rest is left untilled. The system is installed in the fall, in conjunction with the application of anhydrous ammonium. When compared to a no-till method, strip-till can help warm up the seedbed sooner.
- **Ridge-till (ridge-plant or till-plant):** To keep the permanent ridges established for planting a row crop, specialized planters and cultivators must be utilized. At planting, the ridge tops are cleansed of previous crop residue so that the new crop can be planted on the ridges. Crop residue is left undisturbed on the

soil surface after harvest until planting time. A good ridge tillage technique necessitates the maintenance of the ridges, which necessitates the use of adapted or specialized equipment.

- **Mulch-till:** Mulch-till refers to any conservation tillage system other than no-till and ridge-till. Deep tillage may be used, with crop remains mixed up with the soil. Mulch-tilling necessitates the employment of various instruments. The tillage instruments used must leave at least 30% of the residue on the surface of the soil.

Factors to Consider When Selecting the Best Tillage System

- **Weed Management:** Herbicides are used more frequently in conservation tillage techniques to control weeds. In a no-till production system, surface-applied pre-emergence or post-emergence herbicide should be used to control weeds. Herbicide application timing and crop competition enhancement can help manage weeds in a no-till system. To control tiny annual weeds, use paraquat or a Roundup® brand agricultural herbicide. Roundup, unlike paraquat, is successful at controlling perennial weeds because it translocate within the plant. In maize or grain sorghum fields, paraquat or a Roundup brand agricultural herbicide can be tank mixed with atrazine to assist manage perennial grasses like Tall Fescue. For enhanced control of difficult-to-control broadleaves like marestail, combine a Roundup brand agricultural herbicide with 2,4-D or dicamba in a tank mix. Several pre-emergence herbicides that are used to control weeds in conventional tilled fields can also be used in no-till fields, but labels should always be checked for products that can be used in no-till fields. Keep in mind that no-till fields necessitate the higher end of the product rate ranges to guarantee that enough active ingredient is available to break down crop residue and control weeds efficiently. Always read the label for instructions on how to apply the product properly.
- **Organic Matter (OM):** Tillage affects the amount and distribution of organic matter in the soil. The most successful strategy for boosting OM is no-till farming. After several years, the gradual degradation of undisturbed crop residue left on the soil surface aids in the growth of OM in the upper few inches. No-till can also aid increase earthworm populations by increasing root biomass

and microbial diversity. Strip-till and ridge-till systems have an OM level that is halfway between a moldboard plow and a no-till system.

- **Fertilizer Management:** Because the immobile elements of phosphorus (P), potassium (K), and limestone move slowly in most soils unless they are physically combined during tillage operations, tillage techniques have an impact on fertilizer management. Based on soil test results, soil fertility levels for conservation tillage at deeper depths may be reduced, necessitating fertility management.
- **Soil pH and Liming:** Nitrogen (N) fertilizers added to the surface in a no-till system lower the pH of the soil in the top two to three inches. pH samples should be taken from the top two inches of the soil, while nutrient samples should be taken from depths of six to seven inches. To correct soil pH to the right levels for crop growth and development, lime should be added based on soil test recommendations. In fields where triazine herbicides are employed, over-liming or liming right before planting should be avoided. The activity of these herbicides can be increased by a high pH, which could result in crop damage. Half of the prescribed lime should be applied if only the top two to three inches of the soil are acidic.
- **Nitrogen:** To prevent N volatilization losses, liquid N or anhydrous ammonia should be injected into the soil using coulters in front of the injector knives. When injecting nitrogen into four to eight inches below the soil surface in a no-till system, the same rates as for conventional tillage are advised. Use a nitrification inhibitor to prevent N loss if a surface application of N is required in a conservation tillage scheme. In a no-till system, injecting resulted in better corn yields than surface N applications, either broadcast or surface banding, according to research.
- **Phosphorus and Potassium:** If a soil test reveals low P and K levels, using a planter or manure applicator to inject both nutrients is strongly advised.

Contribution to Socio-Economic and Environmental Promotion

- Because there are fewer tillage trips and cultivation operations for seedbed preparation, a reduced tillage system requires less labor time and money. The savings range from \$2.47 per hectare to \$19.13 per hectare (Kimble et al., 2007).

- A great number of studies have calculated the possible savings in fuel costs as a result of reduced tillage. They cost between \$3.58 and \$28.29 per hectare (Kimble et al., 2007).
- Because fewer tillage tools are used, reduced tillage systems have lower machinery repair and maintenance expenses (Kimble et al., 2007).
- Verma and Singh (2009) estimate that zero-tillage technology saves up to US\$70 (Rs. 3200) per hectare in field preparation expenditures, as well as time and labor (up to 10-20 percent). Verma and Singh (2009) found that reducing fuel use by 26.5-43.7 liters per hectare lowers fuel costs and reduces carbon emissions into the atmosphere.
- Due to the mulch on the soil surface, which reduces evapotranspiration, zero-tillage can save farmers roughly 1 million gallons of water per hectare (100mm) compared to conventional techniques (Rehman, 2007).
- Under sub-tropical circumstances, zero-tillage increases soil carbon from 0.1 to 0.7 metric tons ha-1yr-1 (Paustion et al., 1995).

SMART WATER MANAGEMENT

Introduction

Crops are farmed under a variety of water management regimes, ranging from simple soil tillage to sophisticated irrigation technology and management. Around 80% of the world's estimated 1.4 billion ha of cropland is rainfed, accounting for roughly 60% of global agricultural output1. Water management in rainfed situations tries to limit the quantity of water available to a crop by opportunistically diverting the rainwater pathway toward increased moisture storage in the root zone. However, rainfall patterns, not the farmer, determine when and how much water is applied. Irrigation covers over 20% of the world's cropped land and accounts for about 40% of total agricultural output1. This level of output is due to increasing cropping intensities and average yields. Irrigation promotes the concentration of inputs to enhance land production by managing both the amount and timing of water sprayed to crops. Farmers water crops to maintain and enhance yields, as well as to increase the number of crops grown each year. Irrigated yields are two to three times higher than rainfed yields globally. For high-value, high-input cropping systems, a stable and flexible supply of water is essential. However, the economic risk is substantially higher than in rainfed agriculture with smaller inputs. Irrigation can have negative environmental consequences, such as soil salinization and nitrate poisoning of aquifers. Agriculture is under increasing pressure from competing demands for water, as well as environmental imperatives, to produce "more crops with fewer drops" and with less environmental damage. This is a huge concern, and it means that water management for long-term crop intensification will require better, precision agriculture. It will also require agriculture's water management to improve its ability to account for its water consumption in economic, social, and environmental terms. The prospects for long-term intensification vary significantly across different production systems and external demand sources. However, whether rainfed or irrigated, the long-term viability of intensified crop production will be determined by the adoption of ecosystem approaches like conservation agriculture, as well as other key practices like the use of high-yielding varieties and high-quality seeds, and integrated pest management.

Rainfed Cropping Systems

Many rainfed crop varieties have been modified to take advantage of moisture retained in the root zone. Rainfed systems can be improved further by rotating deep-rooting crops, altering crops to create a deeper rooting habit, enhancing soil water storage capacity, improving water infiltration, and decreasing evaporation by organic mulching, among other things. Capturing runoff from nearby fields can further extend the time that soil moisture is available. Improving rainfed agriculture's production is largely dependent on better husbandry in all elements of crop management. Pests and a lack of soil nutrients are more likely to affect productivity than a lack of water. Reduced tillage, organic mulching, and the use of natural and managed biodiversity are all important aspects of better husbandry. As a result, the application of ecosystem-based techniques that maximize moisture storage in the root zone will determine the breadth of SCPI implementation under rainfed conditions. While these methods can help with intensification, the system is still sensitive to rainfall fluctuations. Crop output will be jeopardized as a result of climate change. The task of creating effective climate change adaptation solutions is more pressing in rainfed agriculture than everywhere else. To alleviate farmers' risk aversion, more steps are required. To combat climate change and improve the resilience of production systems, they include better seasonal and yearly rainfall and water supply predictions, as well as flood control. More complex water management actions may be necessary to reduce production risk, but they are not required to increase rainfed productivity. For example, some rainfed farming systems could be converted to low-input supplementary irrigation systems to bridge short dry spells during important growth stages, although this is still dependent on rainfall timing and intensity.

In transitional climates, such as the Mediterranean and portions of the Sahel, on-farm runoff management, including the use of water retention bunds in cultivated areas, has been used successfully to extend soil moisture availability after each rain event. Limited supplementary irrigation can be achieved through off-farm runoff management, such as the concentration of overland flow into shallow groundwater or farmer-managed storage. When implemented across a vast area, however, these interventions have an influence on downstream consumers as well as overall river basin water budgets. The level of agricultural mechanization required to take advantage of rainstorm events will frequently determine how far the beneficial environmental and soil moisture conservation benefits of ecosystem systems may be extended. Simpler approaches, such as opportunistic runoff farming, will always be dangerous, especially in the face of more variable rainfall patterns. They'll also continue to be labor-intensive. At the national level, policymakers will need to precisely quantify the relative contributions of rainfed and irrigated production. If rainfed production may be stabilized by increased soil moisture storage, the physical and socioeconomic conditions that allow this to happen must be well established. The relative merits of low-intensity SCPI investments across large rainfed systems vs high-intensity localized full irrigation investments require rigorous socioeconomic analysis considering development goals. Reorganization and fortification of advising services to farmers depending on rainfed agriculture, as well as fresh attempts to market crop insurance for small-scale producers, are needed in terms of institutions. To stabilize production from existing rainfed systems under climate change impacts, a more detailed examination of rainfall patterns and soil moisture deficiencies would be required.

Irrigated Cropping Systems

Due to double and triple cropping, the real area harvested is expected to be larger than the entire area equipped for irrigation worldwide, which is presently in excess of 300 million ha. Most of the irrigation systems expansion has occurred in Asia, where rice is grown on about 80 million hectares with yields average 5 tons per hectare (compared to 2.3 tons per ha from the 54 million ha of rainfed lowland rice). Irrigated agriculture, on the other hand, is practiced on only 4% of cropped land in Africa, owing to a lack of financial investment. Irrigation is a popular platform for intensification because it provides a place at which inputs can be concentrated. Sustainable intensification, on the other hand, is contingent on the location of water withdrawal and the adoption of SCPI's ecosystem-based measures, such as soil conservation, enhanced varieties, and integrated pest management. The technology used to deliver water, the soil type and slope (most importantly its infiltration characteristic), and the quality of management all affect the uniformity of distribution and irrigation application efficiency.

- **Surface irrigation:** Surface irrigation in the form of a border strip, basin, or furrow is inefficient and inconsistent compared to overhead irrigation (e.g., sprinkler, drip, drip tape).
- **Micro irrigation:** Micro irrigation has been viewed as a technological answer for field irrigation's poor performance as well to save water. Despite significant capital expenses, commercial horticulturists in both rich and underdeveloped countries are increasingly adopting it.
- **Deficit irrigation:** In the commercial cultivation of fruit trees and some field crops that respond positively to controlled water stress at critical growth stages, deficit irrigation and variants such as regulated deficit irrigation (RDI) are gaining traction. Micro-irrigation and "fertigation," in which fertilizers are applied directly to the zone where the majority of the plant's roots develop, are frequently used in conjunction with RDI. In China, the process has been modified to use easier furrow irrigation. The advantages in terms of lower water inputs are obvious, but they will only be realized if the water supply is extremely stable.
- **Precision irrigation:** A crucial platform will be knowledge-based precision irrigation, which will provide farmers with reliable and flexible water application. Both solid set sprinklers and micro-irrigation, which uses soil moisture sensing and crop canopy temperature to specify irrigation depths to be sprayed in different portions of the field, have been tested using automated systems. Precision irrigation and fertilizer delivery via irrigation water are both future possibilities for field crops and horticulture, but they

come with risks. Salt management is a vital aspect in horticulture sustainability, according to recent computer simulations.

Irrigated agriculture has substantial economic implications. Sprinkler and micro-irrigation technologies, as well as the automation of surface irrigation systems, necessitate long-term capital and operational expenditures. Rain guns are one of the most cost-effective capital alternatives for covering wide areas with overhead irrigation, but they have substantial running expenses. Other overhead irrigation methods have large capital costs and are unsuitable for smallholder farming systems without output subsidies. Many public irrigation systems provide subpar service due to design, maintenance, and management flaws. Through institutional reform and the separation of irrigation service supply from larger oversight and regulation of water resources, there is a lot of room for upgrading systems and their administration. Irrigation requires drainage, which is sometimes disregarded, especially where water levels are high and soil salinity is a problem. Drainage investment will be required to improve the productivity and long-term viability of irrigation systems, as well as to ensure proper agricultural input management. Enhanced drainage, on the other hand, increases the likelihood of pollutants being exported, resulting in degradation of streams and aquatic ecosystems. Protected cropping, primarily in shade houses, is gaining popularity in several nations, including China and India, primarily to produce fruits, vegetables, and flowers. Highly intensive closed-cycle production methods, which use conventional irrigation or hydroponic and aeroponic cultures, will become increasingly popular in the long run, particularly in peri-urban areas with strong markets and increasing water constraint. Irrigating with water reduces instream flows, changes their timing, and generates the circumstances for shocks like deadly algal blooms. Salinization and nutrient and pesticide poisoning of water courses and bodies of water are examples of secondary effects. Other environmental trade-offs from irrigated systems include the fact that rice paddies store more organic matter than dry land soils, contribute less nitrate runoff, and emit less nitrous oxide (N2O). Methane and ammonia emissions, which account for 3 to 10% of world emissions, provide a counterbalance. Irrigation systems that are located inside a fully or over-allocated river basin have low efficiency, and crops often consume less than half of the irrigation water they receive. It is vital to distinguish how much water is lost, both beneficially and unproductively, in accounting terms. Irrigation is intended to benefit crop depletion (evapotranspiration): in a perfect world, transpiration would account for all depletion, with 0% evaporation from soil and water surfaces.

Reduced non-productive evaporative losses have the potential to boost water productivity. Water productivity increases at the basin level are focused on minimizing non-beneficial depletion. However, the downstream effects of greater water depletion for agriculture are not neutral: there is evidence of significant reductions in yearly runoff from "improved" higher catchments in peninsular India that have implemented intensive water harvesting. Minimizing nitrogen losses and export from crops requires careful water management. Nitrification is partially halted in freely drained soils, resulting in the release of N2O, whereas ammonium compounds and urea are partially converted to ammonia in saturated (anoxic) conditions, as seen in rice cultivation. As both ammonia and N2O are generated during irrigation's wetting and drying cycles, urea losses might occur in the atmosphere. For root uptake, N must be in nitrate form, yet it can easily move around in solution. For various settings, several protective and slow-release fertilizer compounds are being developed. Phosphate mobilization and transport in drains and waterways is a complicated process. If erosive flow rates are utilized in furrow irrigation, or if sodic soils disperse, phosphorus export from agriculture can occur. Buffer strips at the extremities of fields and along rivers can trap phosphate and, to a lesser extent, nitrate, preventing them from reaching waterways. As a result, a combination of effective irrigation management, tailwater recycling, and phosphate incorporation in the soil can reduce phosphate export from irrigated lands to near zero. Minimizing off-farm externalities, such as salinization and pollutant export, as well as maintaining soil health and growing conditions, are essential for the long-term viability of intensified irrigated agriculture. That should be the primary focus of farm-level practice, technology, and decision-making, and it emphasizes the importance of depletion water accounting and wiser water allocation at basin and catchment scales, as well as a better understanding of the hydrological interactions between different production systems.

Smart Water Management is the Way Forward

Smart water management technologies can make a water delivery system more resilient and efficient, lowering costs and increasing sustainability. Digital meters and sensors, supervisory control, and data acquisition (SCADA) systems, and geographic information systems are examples of high-tech water solutions (GIS). Conventional water and wastewater systems can be transformed into instrumented, interconnected, and intelligent systems using smart technologies.

- **Instrumented:** the ability to detect, sense, measure, and record data.
- **Interconnected:** the ability to communicate and interact with system operators and managers.
- **Intelligent:** the ability to analyze the situation, enable quick responses, and optimize troubleshooting solutions.

Generally, technology for smart water management have four components with a wide range of applications (See table below)

Table 11: Technology for Smart Water Management

Components	**Purpose**	**Example Applications**
1. Digital output instruments (meters and sensors)	To collect and transmit information in real time.	• Rain gauges, flow meters, water quality monitoring and other environmental data • Acoustic devices for real-time leakage detection • Video camera for asset management • Smart water meters for measuring consumption • Pressure monitoring for leakage detection and pump optimization
2. Supervisory control and data acquisition (SCADA) systems	To process information and remotely operate and optimize systems and processes.	• Pressure management • Pump station optimization • Water treatment plant control • Sewage treatment plant control • Environmental controls, reservoirs, flows, etc.
3. Geographic information system (GIS)	To store, manage, manipulate, and analyze spatial information.	• Asset mapping and asset management • Fully integrated network models • Environmental data analysis and management

4. Software	To store, use, and report data. For modelling infrastructure and environmental systems to improve design, decision making, and risk management.	• Usually integrated with GIS and/or SCADA systems to manage water networks, control pressure, monitor leakage, etc. • Improved decision making and risk management • Customer data bases • Smart metering, billing, and collections • Hydraulic design and optimization • Water resources and hydrological modelling for water security • Cloud-based data management and hosting options

Source: Kay, (2011).

Objectives of Smart Water Management

The basic goal of smart water management is to use and recycle water resources in a fair and sustainable manner. Water is becoming an even more valuable resource due to rising population, environmental concerns, and strain on the food and agriculture sectors. Water management technologies and activities aim to achieve the following goals in this regard:

- **Reduce wasting water:** Reduce the amount of water wasted in agriculture, industries, and electricity generation. Precision farming, smart irrigation, agricultural water management, and real-time water metering are examples of high-tech practices that can be implemented. Find out more about the farm software development services we offer.
- **Improve water quality and prevent contamination**: Improve water quality and protect it from contamination from chemical waste and natural pollution like acidification. Companies employ sensors and IoT technologies for real-time monitoring and control to improve and maintain water quality.
- **Enhance the efficiency of water systems:** Water systems such as water collectors, treatment facilities, distribution mains, and wastewater recycling centers can all be made more efficient. Companies may use IoT and data solutions for asset management to maintain

crucial parameters like water pressure, temperature, and flow close at hand, incorporate predictive maintenance, and reduce equipment damage and downtime.

- **Implement leakage control:** Smart water management systems with leak and moisture sensors can be used to control leaks. Leakage control is critical to protect water resources and budgets, as about $3 billion is spent each year to repair the damage caused by leaks.
- **Practice consumption monitoring:** Use IoT-based water management technologies to track your consumption. It aids in the optimization and management of water resources at various levels - families, towns, countries, and the entire globe.

Benefits of Using IoT for Water Management

Water management IoT technologies assist industry stakeholders, governments, and typical consumers in achieving their sustainability and efficiency goals. In this industry, the Internet of Things has already spawned a brand-new concept: The Internet of Water. It entails connecting all the systems and players in the water supply chain, including water sources, treatment plants, and industrial water management systems, distribution facilities, utility and clean energy companies, and consumers, among others, and providing important insights into the state of water resources and equipment used in the supply chain to decision-makers. IoT may be the best technology for equipping the supply chain with the essential processes, data applications, and tools. And here's why:

- **Transparency:** One of the most significant advantages of IoT-based smart water management is the increased transparency of all activities in the water supply chain. Different stakeholders gain valuable insights into their resources and system performance thanks to data collected across the supply chain. As a result, they are better equipped to make educated judgments about how to improve their operations.
- **Immediate response:** Another advantage of combining intelligent water management systems is the capacity to detect or even foresee problems and respond quickly to limit harm. Real-time monitoring of water quality and chemical composition, for example, enables for early detection of even minor pollution and prompt treatment before it becomes problematic.

- **Automation and optimized use of human resources:** Managers can employ IoT water management systems to automate some procedures and maximize the utilization of human resources partially or entirely. The degree and scope of automation varies depending on the industry and individual company requirements. Smart water supply firms and utility networks, for example, can use linked meters, real-time monitoring systems, and dynamic pricing models to automate the full lifecycle of providing water to customers.
- **Optimized cost:** Automation, better human resource utilization, a data-driven strategy, and a proactive approach to equipment maintenance and resource utilization all add up to significant cost savings. One of the reasons why water firms are investigating the use of IoT in water management is to minimize long-term operational costs.
- **Sustainability:** Many retrofit and innovation initiatives, not just in the smart water industry, but in any other area such as energy, construction, logistics, and so on, have sustainability aims at their core. Smart water technologies are no longer viewed solely as a source of cost savings and increased efficiency, but as a means of achieving a variety of environmental goals, including decreased carbon footprint, pollution, and, most importantly, water preservation.
- **Forward-looking strategy:** When it comes to water conservation, smart technology benefits not just businesses and consumers, but also researchers working on long-term water conservation initiatives. Researchers may develop data-driven strategies for optimizing the use of water resources for the benefit of communities, ecosystems, and the entire planet using data provided by water supply management and other connected systems along the supply chain.

INTEGRATED PEST MANAGEMENT

Introduction

IPM is an ecosystem-based strategy that combines a combination of techniques like as biological control, habitat change, cultural practice modification, and the adoption of resistant varieties to prevent pests or their damage over time. Pesticides are only used when monitoring reveals that they are required by defined parameters, and treatments are designed to eliminate just the target organism. Human health, beneficial and nontarget creatures, and the environment are all protected by pest control materials that are carefully chosen and used. IPM, or integrated pest management, is a method for resolving pest problems while reducing threats to people and the environment. IPM can be used to control a variety of pests in a variety of settings, including urban, agricultural, and wildland or natural regions. Increasing your farm's biodiversity through integrated management strategies can have a variety of benefits, including providing beneficial habitats, boosting weed suppression, attracting insects away from your main cash crop, and biochemical pest control.

How Does IPM Work

IPM focuses on long-term prevention of pests or their damage by managing the ecosystem: IPM is taking steps to prevent pests from becoming a problem, such as cultivating a pest-resistant crop, employing disease-resistant plants, or sealing cracks to keep insects and mice out of a structure. Rather of merely removing the pests you're seeing now, IPM requires you to consider the pest's potential to thrive in its habitat. You can create adverse conditions for the pest if you have this information.

In IPM, monitoring and correct pest identification help you decide whether management is needed: Monitoring is inspecting your crop, landscape, forest, building, or other location to determine whether pests are there, how many there are, and how much damage they've caused. Knowing if a pest is likely to become a problem and determining the appropriate management method begins with correctly identifying the pest. You can decide if the pest can be tolerated or whether it is an issue that requires control after monitoring and reviewing information on the pest, its biology, and environmental factors. If you need to exert control, this knowledge can also assist you in determining the most successful management tactics and the optimum moment to employ them.

IPM programs combine management approaches for greater effectiveness: Using a combination of strategies that work better together than separately is the most effective, long-term way to manage pests. Pest management strategies are frequently categorized into the following groups.

- **Biological control:** Biological control is the use of natural enemies to control pests and their damage, such as predators, parasites, viruses, and competitors. Natural enemies include invertebrates, plant diseases, nematodes, weeds, and vertebrates.

- **Cultural controls:** Pest establishment, reproduction, dissemination, and survival are all reduced through cultural restrictions. Changing irrigation procedures, for example, can help reduce insect problems because too much water can promote root disease and weed growth.
- **Mechanical and physical controls:** Mechanical and physical controls kill pests directly, keep them out of the environment, or make it unsuitable for them. Mechanical control includes traps for rats. Mulches for weed control, steam sterilization of the soil for disease control, and barriers like screens to keep birds and insects out are all examples of physical controls.
- **Chemical control:** Pesticides are used in chemical control. Pesticides are only used when necessary and in conjunction with other measures for more effective, long-term control in IPM. Pesticides are chosen and used in such a way that their potential for harm to people, nontarget creatures, and the environment is minimized. You'll use the most selective pesticide that will do the job while also being the safest for other species and air, soil, and water quality; you'll use pesticides at bait stations rather than sprays; and you'll spot-spray a few weeds rather than a whole area with IPM.

These IPM principles and practices are combined to create IPM programs. While each situation is different, six major components are common to all IPM programs:

- Pest identification
- Monitoring and assessing pest numbers and damage
- Guidelines for when management action is needed
- Preventing pest problems
- Using a combination of biological, cultural, physical/mechanical, and chemical management tools
- After action is taken, assessing the effect of pest management

Benefits of IPM

An integrated pest management (IPM) strategy combines preventive and corrective methods to protect pests from generating major issues while posing the least amount of danger or harm to humans and desired environmental components. IPM is a flexible, dynamic strategy that must be updated on a regular basis as new information from management practice results becomes available. IPM programs have an established track record of considerably lowering pesticide-related risks while increasing environmental quality, health, and welfare. The following are some of the advantages of an integrated approach:

- Promotes sound structures and healthy plants
- Promotes sustainable bio-based pest management alternatives.
- Reduces environmental risk associated with pest management by encouraging the adoption of more ecologically benign control tactics
- Reduces the potential for air and ground water contamination
- Protects non-target species through reduced impact of pest management activities.
- Reduces the need for pesticides by using several pest management methods
- Reduces or eliminates issues related to pesticide residue
- Reduces or eliminates re-entry interval restrictions
- Decreases worker, tenant, and public exposure to pesticides
- Alleviates public concern about pest and pesticide-related practices
- Maintains or increases the cost-effectiveness of a pest management program

Adoption of IPM tactics has a positive impact on the economy since it promotes long-term growth, greater productivity, and reduced pest damage. The choices available through IPM eliminate the need for synthetic pesticides entirely. A better environment benefits everyone in the long run.

DIVERSIFY THE SMALLHOLDER FARM

Introduction

Diversified farming systems are a collection of approaches and technologies for producing food in a sustainable manner by harnessing natural diversity at the plot, field, and landscape scales. To support these locally-adapted management systems, DFS rely on a variety of cultures, customs, and governance structures all around the world. DFS allows important ecosystem services – like pollination and pest control – to be generated and regenerated within the agroecosystem, supported by human knowledge to sustain those processes, by sustaining a complex fabric of natural and human ecologies. While there is no single "DFS" design, they all have a focus on local production, agri-ecological and local knowledge, and whole systems approaches to agriculture that promote ecological variety and ecosystem services from the field to the landscape scale. Polycultures (several crops and/or varieties) and the integration of fish or cattle, for example, allow for more efficient fertilizer usage in the crop field. Hedgerows or live fences around the field provide habitat for beneficial insects that suppress pests and pollinate crops. Wild or partly-wild pasture, meadow, woodlots, woods, marshes, streams, and lakes in the surrounding terrain support livestock and wildlife, offer flood control and forest products, and improve water quality. Diversified agricultural systems and agroecological techniques can help rural and urban populations support smallholder livelihoods while also reducing hunger and poverty. Diversified agriculture is essential for feeding the world's population reliably and indefinitely while mitigating climate change and averting the collapse of ecological systems that humans rely on for survival. The new DFS center's ultimate purpose, as a UC Berkeley research hub for investigating multi-scale, systems-based agriculture, is to assist build a resilient, socially just, and secure global food system.

Diversification Types

In many developing countries, the agricultural sector's growth and development is critical for escaping poverty traps. Crop diversity is one of the solutions proposed as a path to poverty alleviation in agriculture, and it is regarded one of the most environmentally feasible, cost-effective, and easy ways to reduce the impact of uncertainties, especially among small-scale farmers. Crop diversification is defined as the addition of new crops to an existing cropping system. Crop diversification can also imply the substitution of high-value commodities for low-value ones, such as fruits and vegetables for export. Diversification can also take the form of combining crops and livestock, which is known as mixed farming, or combining crops and trees, which is known as agroforestry, or producing cash crops (e.g., cashew and coffee). A non-exhaustive overview of the key advantages and main characteristics of different agricultural diversification systems, including mixed farming and agroforestry, is presented in the table below.

Table 12: Diversification of agricultural systems

Diversification types	Description of diversification	Main characteristics
Increased structural diversity	It increases the structural diversity of crops within a field; for example, strip intercropping, which involves growing many crops in strips that are narrow enough for the crops to interact but wide enough to allow autonomous cultivation.	Farm level; same land unit
Genetic diversity in monoculture	Growing mixed varieties of a species in a monoculture.	Farm level; same land unit
High-value crops	A switch from a less profitable and sustainable crop or cropping system to one that is more profitable and sustainable.	Farm level; same land unit
Crop rotations	Temporal diversity through crop rotations.	Farm level; different spaces; different times
Polyculture	Within the same field, grow two or more crop kinds as well as wild types. Crop diversity in terms of space and time.	Farm level; different spaces; different times.
Diversify field with non-crop vegetation	Growing weed strips or vegetation banks in and alongside crops.	Farm level; different spaces
Mixed farming	Crops and livestock.	Farm level; different spaces; different times.
Agroforestry	Growing crops and trees together.	Farm level; different spaces; different times.
Mixed landscapes	Development of larger scale diversified landscapes with multiple ecosystems.	Larger scale; spatial; temporal

Source: Feliciano, (2019).

Advantages of Diversification

In the United States, farms that specialize and expand in size to obtain economies of scale are common. Small and medium-sized farms, on the other hand, require a distinct type of economy. Scope economies rely on managing a variety of businesses in order to gain access to more profitable markets and lower the cost of off-farm inputs. Diversified agricultural systems have been shown to provide ecological and economic benefits in studies, but further research on the process of diversification in US agriculture is needed. The following are some of the benefits of diversification:

- New value chains produce more output.
- New revenue streams are generated by entering new markets.
- Ecosystem services derived from synergies between crop, animal, and farm energy enterprises result in lower production costs.
- Lower cost of capital because human, social, built, and natural capitals can be used to diversify small and medium-sized farms and replace more expensive financial capital.
- For the farm, this means more net revenue and a higher return on investment.

References

1. Bosc, P. M., Berdegué, J., Goïta, M., van der Ploeg, J. D., Sekine, K., & Zhang, L. (2013). Investing in smallholder agriculture for food security (No. 6). HLPE.
2. UNEP, IFAD. (2013). Smallholders, food security and the environment.
3. Christen, R. P., & Anderson, J. (2013). Segmentation of smallholder households: Meeting the range of financial needs in agricultural families (No. 78232, pp. 1-32). The World Bank.
4. Wymann von Dach, S., Romeo, R., Vita, A., Wurzinger, M., & Kohler, T. (2013). Mountain farming is family farming: a contribution from mountain areas to the International Year of Family Farming 2014. FAO.
5. Zimmerer, K. S., Lambin, E. F., & Vanek, S. J. (2018). Smallholder tele coupling and potential sustainability. Ecology and Society, 23(1).
6. Reardon, T., Echeverria, R., Berdegué, J., Minten, B., Liverpool-Tasie, S., Tschirley, D., & Zilberman, D. (2019). Rapid transformation of food systems in developing regions: highlighting the role of agricultural research & innovations. Agricultural systems, 172, 47-59.
7. Elbehri, A., Kaminski, J., Koroma, S., Iafrate, M., & Benali, M. (2013). West Africa food systems: An overview of trends and indicators of demand, supply, and competitiveness of staple food value chains. Rebuilding West Africa's food potential: Policies and private initiatives to promote smallholder-inclusive staple food value chains, 1-42.

8. Dorward, A., Anderson, S., Bernal, Y. N., Vera, E. S., Rushton, J., Pattison, J., & Paz, R. (2009). Hanging in, stepping up and stepping out: livelihood aspirations and strategies of the poor. Development in practice, 19(2), 240-247.
9. Mangnus, E. and Metz, N. (2019). Farmers? Which farmers? Strategies to differentiate categories of farmers as 'target' group of food security interventions. Insight Paper. Utrecht: Utrecht University: Agripofocus.
10. Lowder, S. K., Skoet, J., & Raney, T. (2016). The number, size, and distribution of farms, smallholder farms, and family farms worldwide. World Development, 87, 16-29.
11. Mbow, C., Rosenzweig, C., Barioni, L. G., Benton, T. G., Herrero, M., Krishnapillai, M., ... & Belkacemi, M. (2019). Climate Change and Land: an IPCC special report on climate change, desertification, land degradation, sustainable land management, food security, and greenhouse gas fluxes in terrestrial ecosystems. Nature.
12. Food & agriculture organization. (2018a). Future of food and agriculture 2018: alternative pathways to 2050. FAO.
13. Castaneda, R., Doan, D., Newhouse, D. L., Nguyen, M., Uematsu, H., & Azevedo, J. P. (2016). Who are the poor in the developing world? World Bank Policy Research Working Paper, (7844).
14. Ricciardi, V., Ramankutty, N., Mehrabi, Z., Jarvis, L., & Chookolingo, B. (2018). How much of the world's food do smallholders produce? Global Food Security, 17, 64-72.
15. Ahearn, M. C. (2013). Beginning Farmers and Ranchers at A Glance 2013 Edition (No. 1475-2016-120916).
16. Hoppe, R. A., & Korb, P. J. (2006). Understanding US farm exits (No. 1477-2016-121077).
17. Mueller, S. A., Anderson, J. E., & Wallington, T. J. (2011). Impact of biofuel production and other supply and demand factors on food price increases in 2008. Biomass and bioenergy, 35(5), 1623-1632.
18. Waggle, D., & Johnson, D. T. (2009). An analysis of the impact of timberland, farmland and commercial real estate in the asset allocation decisions of institutional investors. Review of Financial Economics, 18(2), 90-96.
19. Ash, A., McIntosh, P., Cullen, B., Carberry, P., & Smith, M. S. (2007). Constraints and opportunities in applying seasonal climate forecasts in agriculture. Australian Journal of Agricultural Research, 58(10), 952-

965.

20. Pendell, D., Johnson, K., Pritchett, J., Thilmany, D., Seitzinger, A., & Hillberg, A. (2011). Production resources and management: A comparison of low-sales, medium-sales, and large family farm operations in the United States. Available at: http:\\dare. colostate. edu/ pubs/extension. aspx, PFMR, 11-03.
21. Wien, H. C. (2009). Microenvironmental variations within the high tunnel. HortScience, 44(2), 235-238.
22. Martinez, S., Hand, M., Da Pra, M., Pollack, S., Ralston, K., Smith, T., ... & Newman, C. (2010). Local food systems: Concepts, impacts, and issues, ERR 97. US Department of Agriculture, Economic Research Service, 5.
23. Shepherd, M. (2014). Contributing factors to the success of small-scale diversified farms in the Mountain West. Utah State University.
24. Mishra, A. K., El-Osta, H. S., & Shaik, S. (2010). Succession decisions in US family farm businesses. Journal of Agricultural and Resource Economics, 133-152.
25. Thistlethwaite, R. (2012). Farms with a future: Creating and growing a sustainable farm business. Chelsea Green Publishing.
26. Muske, G., Fitzgerald, M. A., Haynes, G., Black, M., Chin, L., MacClure, R., & Mashburn, A. (2009). The intermingling of family and business financial resources: Understanding the copreneurial couple. Journal of Financial Counseling and Planning, 20(2), 27.
27. Yilmazer, T., & Schrank, H. (2010). The use of owner resources in small and family-owned businesses: Literature review and future research directions. Journal of Family and Economic Issues, 31(4), 399-413.
28. Gupta, A. (2004) "The Single Sourcing Decision." ISM Magazine 15(8), 8–9.
29. Wirthgen, A. (2005). Consumer, retailer, and producer assessments of product differentiation according to regional origin and process quality. Agribusiness, 21(2), 191-211.
30. Bennis, W. G., Schein, E. H., Steele, F., & Berlew, D. (1964). Interpersonal Dynamics (Homewood, Ill).
31. Salatin, J. (2007). Everything I want to do is illegal. Polyface, Incorporated.

32. Mojduszka, E. M. (2004). Private and public food safety control mechanisms: Interdependence and effectiveness (No. 377-2016-20799).
33. Whitney, E. (2013). Listeria outbreak still haunts Colorado's cantaloupe growers. National Public Radio, The Salt Whats on Your Plate.
34. Adams, B. (2011, September 3). "Utah peaches coming on strong despite late spring." Salt Lake Tribune. Online. Available at http://archive.sltrib.com/ [Retrieved July 2021].
35. Okurut, N., Schoombee, A., & Van der Berg, S. (2004). Credit Demand and Credit rationing in the Informal financial Sector in Uganda. African Development and Poverty Reduction: The Macro-Micro Linkage. In Forum Paper, Lord Charles Hotel, Somerset West, South Africa.
36. Barrios, S., Ouattara, B., & Strobl, E. (2008). The impact of climatic change on agricultural production: Is it different for Africa? Food policy, 33(4), 287-298.
37. DAFF. (2009). Agricultural education and training access barriers report. Department of Agriculture, Forestry & Fisheries.
38. Dube, E., Chiduza, C., & Muchaonyerwa, P. (2012). Conservation agriculture effects on soil organic matter on a Haplic Cambisol after four years of maize–oat and maize–grazing vetch rotations in South Africa. Soil and Tillage Research, 123, 21-28.
39. Hall, R., & Aliber, M. (2010). The case for re-strategising spending priorities to support small-scale farmers in South Africa (p. 20).
40. Kassam, A. H., Friedrich, T., & Derpsch, R. (2010, October). Conservation agriculture in the 21st century: A paradigm of sustainable agriculture. In European Congress on Conservation Agriculture (Vol. 10, pp. 4-6).
41. Kurukulasuriya, P., & Rosenthal, S. (2013). Climate change and agriculture: A review of impacts and adaptations.
42. OECD. 2006. Agricultural Policy Reform in South Africa. Policy Briefs. Organization for Economic Co-operation and Development. Available at www.oecd.org/publications/ policybriefs
43. Femi, O. A., & van Schalkwyk, H. D. (2006). Mentorship Alliance between South African Farmers: Implication for Sustainable Agriculture Sector Reform (No. 1004-2016-78492).

44. Ortmann, G. F., & King, R. P. (2006). Small-scale farmers in South Africa: Can agricultural cooperatives facilitate access to input and product markets? (No. 1701-2016-139140).
45. Alston, J. M., & Martin, W. J. (1995). Reversal of fortune: immiserizing technical change in Agriculture. American Journal of Agricultural Economics, 77(2), 251-259.
46. Essiet, E. U. (1990). A comparison of soil degradation under smallholder farming and large-scale irrigation land use in Kano State, Northern Nigeria. Land Degradation & Development, 2(3), 209-214.
47. Kusterer, D. G. A. K., & Glover, K. (1990). Small Farmers, Big Business.
48. Tisdell, C. A. (Ed.). (2005). Economics of environmental conservation (Vol. 1). Edward Elgar Publishing.
49. U.S. Department of Agriculture, Economic Research Service, (1992). Economic Indicators of the Farm Sector: National Financial Summary, 1991. USD.MERS, Washington DC.
50. Reardon, T., Echeverria, R., Berdegué, J., Minten, B., Liverpool-Tasie, S., Tschirley, D., & Zilberman, D. (2019). Rapid transformation of food systems in developing regions: highlighting the role of agricultural research & innovations. Agricultural systems, 172, 47-59.
51. Haggblade, S., Hazell, P., & Reardon, T. (2010). The rural non-farm economy: Prospects for growth and poverty reduction. World development, 38(10), 1429-1441.
52. Reardon, T., & Timmer, C. P. (2014). Five inter-linked transformations in the Asian agrifood economy: Food security implications. Global Food Security, 3(2), 108-117.
53. Lowder, S. K., Skoet, J., & Raney, T. (2016). The number, size, and distribution of farms, smallholder farms, and family farms worldwide. World Development, 87, 16-29.
54. Jayne, T. S., Chamberlin, J., Traub, L., Sitko, N., Muyanga, M., Yeboah, F. K., ... & Kachule, R. (2016). Africa's changing farm size distribution patterns: the rise of medium-scale farms. Agricultural Economics, 47(S1), 197-214.
55. Jayne, T. S., Muyanga, M., Wineman, A., Ghebru, H., Stevens, C., Stickler, M., ... & Nyange, D. (2019). Are medium-scale farms driving agricultural transformation in sub-Saharan Africa? Agricultural Economics, 50, 75-95.

56. Burke, W. J., Jayne, T. S., & Sitko, N. J. (2020). Do medium-scale farms improve market access conditions for Zambian smallholders? Journal of Agricultural Economics, 71(2), 517-533.

57. Chamberlin, J., & Jayne, T. S. (2020). Does farm structure affect rural household incomes? Evidence from Tanzania. Food Policy, 90, 101805.

58. Gneiting, U., & Sonenshine, J. (2018). A living income for small-scale farmers: Tackling unequal risks and market power.

59. FAO, 2018a. The future of food and agriculture – alternative pathways to 2050. Rome: FAO. 64 pages. Licence: CC BY-NC-SA 3.0 IGO. ISBN 978- 92-5-130989-6.

60. World Economic Forum, 2017. Shaping the future of global food systems: A scenarios analysis. Davos: World Economic Forum.

61. Searchinger, T., Waite, R., Hanson, C., Ranganathan, J., Dumas, P., Matthews, E., & Klirs, C. (2019). Creating a sustainable food future: a menu of solutions to feed nearly 10 billion people by 2050. Final report. WRI.

62. Inter-Agency Research and Analysis Network (IARAN). 2019. Global food systems: an outlook to 2050. Paris: Institut de Relations Internationales et Stratégiques (IRIS).

63. Birner, R. 2018. Fostering political will to drive agricultural transformation. In: Africa agriculture status report: catalyzing government capacity to drive agricultural transformation (Issue 6) Sumba, D. [ed]. Nairobi, Kenya: AGRA.

64. Woodhill, J. (2010). Capacities for institutional innovation: a complexity perspective. IDS bulletin, 41(3), 47-59.

65. Vorley, W. (2002). Sustaining Agriculture: Policy, Governance, and the Future of Family-based Farming: a Synthesis Report of the Collaborative Research Project'policies' that Work for Sustainable Agriculture and Regenerating Rural Livelihoods'. IIED.

66. Hazell, P. B., & Rahman, A. (Eds.). (2014). New directions for smallholder agriculture. OUP Oxford.

67. Wiggins, S. (2014). Rural non-farm economy: current understandings, policy options, and future possibilities. New Directions for Smallholder Agriculture. Oxford University Press, Oxford, 1-46.

68. Mellor, J. W. (2017). Agricultural development and economic transformation: promoting growth with poverty reduction. Springer.

69. Woodhill, J. (2016). Inclusive Agribusiness: The State of Play

Background Working Paper. Global Donal Platform for Rural Development.

70. Woodhill, J., Hasnain, S., & Griffith, A. (2020). Farmers and Food Systems: What Future for Small-scale Agriculture. Environmental Change Institute, University of Oxford: Oxford, UK.

71. Johnston, B. F., & Kilby, P. (1975). Agriculture and structural transformation; economic strategies in late-developing countries. Agriculture and structural transformation; economic strategies in late-developing countries.

72. Mellor, J. W. (1976). The new economics of growth; a strategy for India and the developing world.

73. Gereffi, G., & Lee, J. (2012). Why the world suddenly cares about global supply chains. Journal of supply chain management, 48(3), 24-32.

74. USAID. 2012. "Value Chain Development." MicroLINKS wiki (http://apps.develebridge.net/amap/index.php/Value_Chain_Development).

75. Csaki, C., & de Haan, C. (2003). Reaching the rural poor: a renewed strategy for rural development. World Bank Publications.

76. IFAD, U. (2013). Smallholders, food security and the environment. Rome: International Fund for Agricultural Development, 29.

77. Nagayets, O. (2005). Small farms: current status and key trends. The future of small farms, 355.

78. Feder, G. (1985). The relation between farm size and farm productivity: The role of family labor, supervision and credit constraints. Journal of development economics, 18(2-3), 297-313.

79. Singh, I., Squire, L., & Strauss, J. (1986). Agricultural household models: Extensions, applications, and policy.

80. Chen, M. A., & Dunn, E. (1996). Household economic portfolios.

81. Jayne, T. S., Mather, D., & Mghenyi, E. (2010). Principal challenges confronting smallholder agriculture in sub-Saharan Africa. World development, 38(10), 1384-1398.

82. Fowler, B., & Brand, M. (2011). Pathways out of poverty: Applying key principles of the value chain approach to reach the very poor. USAID, Microreport, 173.

83. Garloch, A. C. (2012). Pushing the Poverty Frontiers of Inclusive Value Chain Development.

84. Dunn, E., Schiff, H., & Creevey, L. (2011). Linking small-scale vegetable farmers to supermarkets: effectiveness assessment of the GMED India project. USAID AMAP microReport, 166.
85. Dunn, E. (2014). Smallholders and inclusive growth in agricultural value chains.
86. Bardhan, P. (1991). The economic theory of agrarian institutions.
87. Dries, L., Germenji, E., Noev, N., & Swinnen, J. F. (2009). Farmers, vertical coordination, and the restructuring of dairy supply chains in Central and Eastern Europe. World development, 37(11), 1742-1758.
88. Henson, S., Masakure, O., & Boselie, D. (2005). Private food safety and quality standards for fresh produce exporters: The case of Hortico Agrisystems, Zimbabwe. Food policy, 30(4), 371-384.
89. Maertens, M., & Swinnen, J. F. (2009). Trade, standards, and poverty: Evidence from Senegal. World development, 37(1), 161-178.
90. Sadler, M. (2006). Vertical coordination in the cotton supply chains in Central Asia. The Dynamics of Vertical Coordination in Agrifood Chains in Eastern Europe and Central Asia: Case Studies, 73-114.
91. Dries, L., Germenji, E., Noev, N., & Swinnen, J. F. (2009). Farmers, vertical coordination, and the restructuring of dairy supply chains in Central and Eastern Europe. World development, 37(11), 1742-1758.
92. Berkum, S. V. (2006). Vertical coordination in the dairy sector: a comparative analysis between Romania and Slovakia.
93. Van Herck, K., Swinnen, J. F., & Deconinck, K. (2012). How the east was won: Supply chain restructuring in the Eastern European beer market. German Journal of Agricultural Economics, 61(670-2016-45941), 213-222.
94. Dolan, C., & Humphrey, J. (2000). Governance and trade in fresh vegetables: the impact of UK supermarkets on the African horticulture industry. Journal of development studies, 37(2), 147-176.
95. Gow, H. R., Streeter, D. H., & Swinnen, J. F. (2000). How private contract enforcement mechanisms can succeed where public institutions. Agricultural economics, 23(3), 253-265.
96. Reardon, T., Vrabec, G., Karakas, D., & Fritsch, C. (2003). The Rapid Rise of Supermarkets in Croatia: implications for farm sector development and agribusiness competitiveness programs. Michigan State University.

97. Kolavalli, S., Mensah-Bonsu, A., & Zaman, S. (2015). Agricultural value chain development in practice: private sector-led smallholder development.
98. World Bank. (2011). The Dynamics of Vertical Coordination in Agrifood Chains in Eastern Europe and Central Asia: Implications for Policy and World Bank Operations. World Bank.
99. Serova, E., & Karlova, N. (2010). The Russian Federation: review of the dairy sector. FAO Investment Centre. Country Highlights (FAO).
100. Williamson, O. E. (2008). The economic institutions of capitalism. NH Barma, & SK Vogel, The Political Economy Reader: Markets as Institutions, 185-195.
101. GEREFFI, G., HUMPHREY, J., & STURGEON, T. (2008). The Governance of Global Value Chains.
102. Maertens, M., Colen, L., & Swinnen, J. F. (2011). Globalisation and poverty in Senegal: a worst-case scenario? European Review of Agricultural Economics, 38(1), 31-54.
103. Suzuki, A., Jarvis, L. S., & Sexton, R. J. (2011). Partial vertical integration, risk shifting, and product rejection in the high-value export supply chain: The Ghana pineapple sector. World Development, 39(9), 1611-1623.
104. Swinnen, J. F., & Maertens, M. (2007). Globalization, privatization, and vertical coordination in food value chains in developing and transition countries. Agricultural economics, 37, 89-102.
105. Gataulina, Ekaterina A., Vasilii Y. Uzun, Alexander V. Petrikov, and Renata G. Yanbykh. 2006. "Vertical Integration in an Agroindustrial Complex: Agrofirms and Agroholdings in Russia." In The Dynamics of Vertical Coordination in Agrifood Chains in Eastern Europe and Central Asia: Case Studies, edited by Johan Swinnen, 45–71.
106. Miyata, S., Minot, N., & Hu, D. (2009). Impact of contract farming on income: linking small farmers, packers, and supermarkets in China. World development, 37(11), 1781-1790.
107. Key, N., & Runsten, D. (1999). Contract farming, smallholders, and rural development in Latin America: the organization of agroprocessing firms and the scale of outgrower production. World development, 27(2), 381-401.

108. Minot, N., & Ngigi, M. (2004). Are horticultural exports a replicable success story? evidence from Kenya and Côte d'Ivoire. Intl Food Policy Res Inst.

109. Subervie, Julie, and Isabelle Vagneron. (2013). "A Drop of Water in the Indian Ocean? The Impact of GlobalGap Certification on Lychee Farmers in Madagascar." World Development 50:57–73.

110. Schuster, M., & Maertens, M. (2013). Do private standards create exclusive supply chains? New evidence from the Peruvian asparagus export sector. Food Policy, 43, 291-305.

111. Handschuch, C., Wollni, M., & Villalobos, P. (2013). Adoption of food safety and quality standards among Chilean raspberry producers–Do smallholders benefit? Food Policy, 40, 64-73.

112. Kersting, S., & Wollni, M. (2012). New institutional arrangements and standard adoption: Evidence from small-scale fruit and vegetable farmers in Thailand. Food policy, 37(4), 452-462.

113. Gulati, A., Minot, N., Delgado, C., & Bora, S. (2007). Growth in high-value agriculture in Asia and the emergence of vertical links with farmers. Global supply chains, standards and the poor: How the globalization of food systems and standards affects rural development and poverty, 91-108.

114. Vandemoortele, T., Rozelle, S., Swinnen, J., & Xiang, T. (2012). Quality and inclusion of producers in value chains: a theoretical note. Review of Development Economics, 16(1), 122-136.

115. Warning, M., & Key, N. (2002). The social performance and distributional consequences of contract farming: An equilibrium analysis of the Arachide de Bouche program in Senegal. World Development, 30(2), 255-263.

116. Andersson, C. I., Chege, C. G., Rao, E. J., & Qaim, M. (2015). Following up on smallholder farmers and supermarkets in Kenya. American journal of agricultural economics, 97(4), 1247-1266.

117. Briones, R. M. (2015). Small farmers in high-value chains: binding or relaxing constraints to inclusive growth? World Development, 72, 43-52.

118. Bolwig, S., Gibbon, P., & Jones, S. (2009). The economics of smallholder organic contract farming in tropical Africa. World Development, 37(6), 1094-1104.

119. Hansen, H., & Trifković, N. (2014). Food standards are good–for middle-class farmers. World Development, 56, 226-242.

120. Narayanan, S. (2014). Profits from participation in high value agriculture: Evidence of heterogeneous benefits in contract farming schemes in Southern India. Food Policy, 44, 142-157.

121. Setboonsarng, S., Leung, P., & Stefan, A. (2008). Rice contract farming in Lao PDR: Moving from subsistence to commercial agriculture.

122. Ton, Giel, Wytse Vellema, Sam Desiere, Sophia Weituschat, and Marijke D'Haese. 2016. "Effectiveness of Contract Farming for Improving Income of Smallholder Farmers-Preliminary Results of a Systematic Review." http://ageconsearch.umn.edu/bitstream/244

123. Bellemare, M. F., & Novak, L. (2017). Contract farming and food security. American Journal of Agricultural Economics, 99(2), 357-378.

124. Chege, C. G., Andersson, C. I., & Qaim, M. (2015). Impacts of supermarkets on farm household nutrition in Kenya. World Development, 72, 394-407.

125. Asfaw, S., Mithöfer, D., & Waibel, H. (2010). Agrifood supply chain, private-sector standards, and farmers' health: evidence from Kenya. Agricultural Economics, 41(3-4), 251-263.

126. Swinnen, J., & Vandeplas, A. (2012). Rich consumers and poor producers: quality and rent distribution in global value chains. Journal of Globalization and Development, 2(2).

127. Xiang, T., Huang, J., Kancs, D. A., Rozelle, S., & Swinnen, J. (2012). Food standards and welfare: General equilibrium effects. Journal of Agricultural Economics, 63(2), 223-244.

128. Mano, Y., Yamano, T., Suzuki, A., & Matsumoto, T. (2011). Local and personal networks in employment and the development of labor markets: Evidence from the cut flower industry in Ethiopia. World Development, 39(10), 1760-1770.

129. Webber, C. M., & Labaste, P. (2009). Building competitiveness in Africa's agriculture: a guide to value chain concepts and applications. World Bank Publications.

130. Van den Broeck, G., Swinnen, J., & Maertens, M. (2017). Global value chains, large-scale farming, and poverty: Long-term effects in Senegal. Food policy, 66, 97-107.

131. Barrientos, S., Dolan, C., & Tallontire, A. (2003). A gendered value chain approach to codes of conduct in African horticulture. World development, 31(9), 1511-1526.
132. Colen, L., Maertens, M., & Swinnen, J. (2012). Private standards, trade and poverty: GlobalGAP and horticultural employment in Senegal. The World Economy, 35(8), 1073-1088.
133. Maertens, M., & Verhofstadt, E. (2013). Horticultural exports, female wage employment and primary school enrolment: Theory and evidence from Senegal. Food Policy, 43, 118-131.
134. Moustier, P., Tam, P. T. G., Anh, D. T., Binh, V. T., & Loc, N. T. T. (2010). The role of farmer organizations in supplying supermarkets with quality food in Vietnam. Food policy, 35(1), 69-78.
135. Chagwiza, C., Muradian, R., & Ruben, R. (2016). Cooperative membership and dairy performance among smallholders in Ethiopia. Food Policy, 59, 165-173.
136. Fischer, E., & Qaim, M. (2012). Linking smallholders to markets: determinants and impacts of farmer collective action in Kenya. World development, 40(6), 1255-1268.
137. Ito, J., Bao, Z., & Su, Q. (2012). Distributional effects of agricultural cooperatives in China: Exclusion of smallholders and potential gains on participation. Food policy, 37(6), 700-709.
138. Omta, S. W. F., van Elzakker, B., & Schoenmakers, W. W. M. E. (2014). IOB Review-Riding the wave of sustainable commodity sourcing. Review of the Sustainable Trade Initiative IDH 2008-2013 (No. 397). Ministry of Foreign Affairs.
139. Devaux, A., Horton, D., Velasco, C., Thiele, G., López, G., Bernet, T., ... & Ordinola, M. (2009). Collective action for market chain innovation in the Andes. Food policy, 34(1), 31-38.
140. Kolavalli, S., Mensah-Bonsu, A., & Zaman, S. (2015). Agricultural value chain development in practice: private sector-led smallholder development.
141. Marshall, E., & Nair, N. G. (2009). Make money by growing mushrooms. Food and Agriculture Organization of the United Nations (FAO).
142. MDA(2011): Conservation Practices, Minnesota Conservation Funding Guide, Minnesota Department of Agriculture. Available at: http://www.mda.state.mn.us/protecting/conservation/practices/constillage.

143. Naylor, R.L., Hardy, R.W., Buschmann, A.H. et al. A 20-year retrospective review of global aquaculture. Nature 591, 551–563 (2021). https://doi.org/10.1038/s41586-021-03308-6

144. CIMMYT (2010): Resource conserving technologies in South Asia: Frequently asked question. Jat ML, Singh RG, Sidhu HS, Singh UP, Malik RK, Kamboj BR, Jat RK, Singh V, Hussain I, Mazid MA, Sherchan DP, Khan A, Singh VP, Patil SG, Gupta R. pp 1-32.

145. Kimble, J. M., Rice, C. W., Reed, D., Mooney, S., Follett, R. F., & Lal, R. (2007). Soil carbon management: economic, environmental and societal benefits. CRC press.

146. Verma, M. P., & Singh, J. P. (2009). Zero Tillage Technology is an Alternate Method of Sowing-a case study. Quoted from site www. scribd. com.

147. Rehman, A. (2007). Zero tillage technology for rice and wheat crops. Quoted from site www. archives. dawn. com.

148. Paustian, K., Cole, C. V., Sauerbeck, D., & Sampson, N. (1998). CO2 mitigation by agriculture: an overview. Climatic change, 40(1), 135-162.

149. Kay, M. (2011). Water for food: innovative water management technologies for food security and poverty alleviation. UNCTAD/DTL/STICT/2011/2, United Nations.

150. Carballo, E., Eer, A. V., Chie, T. V., & Hilbrands, A. (2008). Small-scale freshwater fish farming. Agromisa/CTA.

151. Feliciano, D. (2019). A review on the contribution of crop diversification to Sustainable Development Goal 1 "No poverty" in different world regions. Sustainable Development, 27(4), 795-808.

INDEX

A

Accelerated Microenterprise Advancement Project 39
Advertising 18
Aeroponic cultures 219
Agribusinesses 6
Agricultural economy 14
Agricultural enterprises 2
Agricultural infrastructure 12, 13
Agricultural labor 50
Agricultural markets 25
Agricultural policies 33
Agricultural resources 38, 50
Agricultural skills 14
Agricultural yield 65
Agritourism 20
Agrochemicals 29
Agroforestry 189, 229
Agronomic skills 62
Air pollution 211
Ammonia toxicity 163
Animal feed 44, 45
Animal husbandry 7, 155
Anthropology 193
Aquaculture 199, 200, 201, 202, 203, 204, 206
Aquaponics 199, 201, 209
Aquatic animals 199, 201, 204, 206
Artificial fertilizers 194
Artificial insemination (AI) 119

B

Beekeeping 144, 145, 146, 147, 150, 151, 152
Beekeeping business 145, 146, 150, 151
Biochemistry 193
Biodiversity 9, 10, 28
Biological control 225

Biomass 199, 207, 213
Business Development Services 39
Business environment 32, 33, 34
Business model 115, 116
Business models 53

C

Calf milk replacers (CMRs) 48
Carbohydrates 90, 173
Chemical fertilizers 160
Chemical pesticides 187
Chicken husbandry 123
Conservation agriculture (CA) 186
Corporate farming 3
Credit systems 6
Crop-livestock systems 189
Cropping systems 50
Crop rotation 8

D

Dairy farmers 118
Dairy industry 18
Department of Agriculture, Forestry, and Fisheries (DAFF) 14
Direct marketing 124
Disease risks 25
Domestic consumers 40
Domestic labor resources 52
Domestic manufacturers 30

E

Economic growth 14, 30, 38, 39, 61
Economic pyramid 22, 23
Ecosystem resource 12
Egg production 125, 127, 132, 138, 139
Environmental quality 185, 227
Environmental sensitivity 9

F

Farm equipment 180
Farming profit margins 180
Farm investment 68
Farm machinery 14
Farm manager 18
Feeding systems 137
fertilizer deep placement (FDP) 186
Fertilizer management 214
Financial planning 17
Financial services 32
Food and Agriculture Organization 25
Food production 2, 8
Food sector entrepreneurs 22
Food security 2, 4, 14, 26, 30, 34
Foreign business 22
Foreign direct investment 49, 69
Four-wheel tractors (4WT) 183

G

Geographic information systems 220
Global agribusiness 23
Global Agriculture and Food Security Program (GAFSP) 64
Global aquaculture production 155
Global fragmentation 26
Gobal warming 13
Gooseberries 110, 114, 115
Greenhouse gases 199
Green Revolution 183, 186
Growing systems 95

H

Healthcare 23
Herbicides 187, 190, 194, 209, 212, 213, 214
Horticultural export enterprises 45

Horticulture 43
Horticulture crops 50
Human capital 14
Human development 14
Hydroponic plant 208
Hydroponics 199, 201, 202, 203, 204, 206

I

Industrial agriculture 2
Integrated-Aqua-Vegeculture-System (iAVs) 200
Integrated farming 193
Integrated pest management 63
Intensive systems 123
Interagency Research and Analysis Network (IARAN) 25
Interlinking markets 68
International Finance Corporation's (IFC) 64
Internet sales 18
Irrigation systems 13
Irrigation technology 215

L

Labor and management 50
Land fragmentation 51
Land management 189, 190
Low-risk portfolio 16

M

Manure control 141
Market development 13, 22
Market potential 165
Market price 25
Market vendors 42
Medicinal plants 194
Meteorology 193
Metric system 129
Microbial growth 83
Micro-dairy 115, 116
Microenterprises 38, 39, 42
Microgreens 73, 74, 75, 77, 78, 79, 81, 83, 84, 88
Microgreens business 75, 76, 79
Micro-irrigation 218
Microorganisms 180, 187
Moisture management 83
Mushroom reproductive cycle 92

N

Natural disasters 25, 28, 30
Natural resource 14
Non-governmental organization (NGO) 64
Nutrient Film Technique (NFT) 202
Nutrition security 4, 28, 33

O

Organic farming 193
Organic fertilizer 8
Organizing agricultural 3

P

Partial mixed ration (PMR) 120
Perennials 197
Periphyton 156, 161, 162, 163
Permaculture 190, 191, 193, 194
Pesticides 190, 209, 226, 227
Phytoplankton 162
Plant growth 197, 201
Plant propagation 181
Polyculture 196, 200, 209
Pond systems 156
Poultry manure 142
Public investment 27

Q

Quality control 44, 59, 60

R

Regulated deficit irrigation (RDI) 218
Restaurants 75, 77, 78, 79, 124, 145
Risk management 49, 51, 53
Road transportation system 13
Robotic systems 120
Rriangle system 67
Rural road network 13

S

Scavenger feed resource base (SFRB) 143
Scenario analysis 25
Shrubs 195, 197, 199
Small Business Administration (SBA) 18
Small businesses 12
Small farm operators 6
Small-scale agriculture 2, 25, 26, 27, 28, 30, 31, 34
Small-scale farmers 2, 15, 26, 27, 28, 29
Small-Scale Integration 69
Snail farming 164, 165, 175, 176
Social protection systems 33
Socioeconomic 15, 26
Sociology 193
Soil cultivation 210
Soil erosion 181, 210, 211, 212
Soil health 8
Soil management 211
Soil moisture 189, 211, 212, 216, 217, 218
Soil organisms 186, 187
Soil preparation 181
Soil quality 8, 25
Soil systems 186
Special purpose vehicles 47, 67
Supervisory control, and data acquisition (SCADA) 220
Sustainable agricultural development 9
Sustainable Development 2
Sustainable farming 7
Sustainable food systems 2

T

Topography 158
Total mixed ration (TMR) 120
Transportation systems 13
Tropical mushrooms 96
Two-wheel tractors (2WT) 183

U

United States Agency for International Development (USAID) 39
United States Department of Agriculture (USDA) 170

V

value chain development (VCD) 61
Ventilation system 136

W

Water management 215, 222, 223
Water quality 17, 187, 221, 222, 223, 226, 228
Weed control 181, 212, 226
Welfare assessments 58
Wholesalers 40, 43, 46, 60
Wimm Bill Dann (WBD) 48
World Economic Forum (WEF) 25
World food supply 4

Printed in the United States
by Baker & Taylor Publisher Services